THEY FLEW
INTO
OBLIVION

They Flew
into
Oblivion

The Disappearance of Flight 19

By

Gian J. Quasar

A true story of Mystery, Irony, and Infrared

Brodwyn-Moor & Doane
2013

Library of Congress Cataloging in Publication Data

Quasar, Gian Julius 1966-

They Flew into Oblivion—
The disappearance of Flight 19

1ˢᵗ Edition

ISBN 978-0-9888505-0-7

Bibliography
1. Flight 19   I. Title

Introduction

In 2005 the national spotlight was placed upon a mystery of aviation because of the story you are about to read. Having an early and still-unfinished copy of this manuscript, Larry Landsman, SCI-FI Channel's determined and slightly maverick Special Projects Director, pushed for a two-hour documentary to be produced by NBC News Productions and then lobbied Congress through Podesta-Matoon, the nation's third largest lobbyist, for formal recognition of the subject of this book in Congress. This culminated in a Resolution in Congress sponsored by Republican E. Clay Shaw of Florida, which passed overwhelmingly at 420-2 votes.

This recognition was unique in that it honored 14 US Navy airmen who had *vanished* 60 years before on December 5, 1945. Yet they were not war heroes. Nor were they on some crucial mission. World War II had been over for months, and the men were on but a routine training mission. They stand out in history because of their number, and the real facts were lost to history for the same reason. The disappearance of the "Lost Patrol," "Lost Squadron" or, as it is more commonly known, Flight 19, captured the imagination of the nation like no other mystery. It is a unique moment in time when 5 aircraft simultaneously and utterly vanish from the earth. It was by all acceptable standards an "impossible" disappearance. It is time that the actual facts are brought to light, mysteries both accepted and challenged. Only from this point can a solution be found even 65 years after they flew into oblivion.

Contents

Prologue

Five aircraft simply do not disappear while in routine formation and leave no trace whatsoever. There has to be an explanation somewhere for something like this. Such an event simply cannot be dismissed after routine investigation and filed away. And yet for the last 68 years this has been just the case for those 5 Avenger torpedo bombers designated Flight 19. James P. O'Donnell once wrote: "The gap between the generations, unless it is to become a chasm and make history meaningless, must somehow be bridged. Each generation owes an after-action report to the generation that follows." Yet for this very unusual mystery there has been no accountability. Perhaps it is not an earth-shaking historical event. But it is nonetheless a world famous incident that continues to elicit questions that have not been answered. It has, in truth, been bridged for the last 68 years by the fabulous and, more often than not, the ludicrous.

According to one popular theory UFOs may have sucked up the flight and spirited it away; in another, time warps may have sent it through a "tear in the curtain of time." But the truth of Flight 19 is only hidden by the funhouse. It stands alone and reluctantly aloof from sober theorizing for 3 vital reasons. The first: sheer number. The disappearance of five aircraft is not like the disappearance of one aircraft. It cannot be compared to Amelia Earhart disappearing between far-flung ports over the Pacific in her Lockheed Electra in 1937, nor to Glenn Miller disappearing in a war zone over the English Channel in a light reconnaissance aircraft in 1944. Five, by its very number, begs many more questions whose answers must, no matter how we might disapprove, flirt with the fantastic.

The second: witnesses and evidence. To an extent, Flight 19 exists in a blurry firmament between both. It had neither and yet had too much. It did not simply vanish, leaving us with no clue. Garbled dialogue and fragmented sentences were picked up for over two and a half hours by base stations, lending fuel to much speculation and embroidery. Men thus partook in the drama, but they were not witnesses. Their observations were recorded in a report over 500 pages in length wherein the Board of Inquiry merely tried to place in order contradictory testimony. The third vital reason is secrecy. Of this it had too much. The evidence their collective testimony could have provided was silenced. The report on Flight 19 was unnecessarily kept restricted; it was not generally released until 30 years later under the Freedom of Information Act.

That gulf is vast and densely populated and given its depth only with popular ideas; some good, some sensational, some bad and some banal. That gulf was also the medium in which spawned the enigma of the Bermuda Triangle, and Flight 19 became its most famous case. This indeed buried it in a vast vault. Any recounting of the incident thereafter was but a vignette designed to link it with the many other aircraft and ships that had vanished.

After its release the Board of Inquiry report helps to vanquish some popular notions, but in some instances it expands rather than takes away from the final enigma of the flight. For example, a study of its loosely compiled radio logs and dispatches prove that the carrier *Solomons* was in a position at the time the flight ran out of fuel to have detected them at sea with its radar. Yet it did not.

The example of the *Solomons* wields a double-edge sword, for it disproves both the banal and the sensationalistic. It disproves the mundane Navy notion that the flight merely ditched at sea, and it proves the flight did not disappear in the Bermuda Triangle.

However, far more tragically, the example of the *Solomons* also mutely testifies to the paucity of the Board's analysis of the evidence. The fact such a relevant clue as the *Solomons'* position could be overlooked reveals the ennui of the Board's own investigative curiosity. For perhaps what is the greatest mystery of aviation we are thus left with only imperfectly known circumstances, without true witnesses, without analysis, not surprisingly without official solutions. In essence, the circumstances were left bundled up, a wealth of dots left unconnected.

But I am grateful to the report and to the officers who compiled it. Even though they were often unforgivably obtuse, they did fulfill what they believed was their primary order to "set in place the circumstances," as imperfectly as they were known. However, a genuine study of what they carelessly put together— radio logs, dispatches, radar reports— makes it possible to connect the dots and go beyond 1945 and solve much of Flight 19's mystery today.

Without this volume you are about to read those decades since would truly have become that chasm impassible and Flight 19 would have forever remained in the pulp ether. The end of the journey must now be decomposed metal, faded paint, and no more human remains. But the chasm of time is not so great as to render the journey meaningless. Although many involved in the drama of that confusing night have since passed, some still remain, each trying to understand how 5 aircraft and 14 young men could vanish in one night. And because of Flight 19's fame, millions have come to share their desire.

For these very reasons Flight 19 is not some irrelevant historical sideshow or common pulp mystery. Yearly, old comrades come to Fort Lauderdale to remember them in a memorial service. I know of no such event where aged groups gather to mourn over the *Hindenburg*'s loss. A disappearance is different, especially one so odd and full of acrimony and unanswered questions. Those who come perhaps are seeking closure to an incident which never should have happened and yet did, and then should have been answered and explained and yet never was.

As incredible as it may seem, the last true words coming from any member of Flight 19 may have been 21 days after the flight vanished. They were in a telegram supposedly sent by one of the crewmen to his brother. "You have been misinformed about me. Am very much alive." The telegram is real; the signature was a family nickname unknown except by the actual family. That telegram is, like Flight 19, an enigma. If that was from crewman George Paonessa, he spoke words that have in essence become those suitable for the epitaph of Flight

19. This incident still calls out these words more poignantly than the last ones popularly attributed to the flight: "I am lost!"

This telegram, received by Corporal Joseph Paonessa at his barracks in Jacksonville, Florida, on the day after Christmas 1945, capped off a year that is now being regarded as the most pivotal in the 20th century. In 1945 sweeping news arrived day after day about the great advances made on the warfront. Then President Roosevelt's sudden death was announced. Germany surrendered. Hitler had committed suicide. The atomic bomb was dropped and mankind faced the power of the atom and a new and uncertain postwar world. It was indeed the year of sweeping changes. When news came over that an entire squadron of TBM Avengers, large Navy torpedo bombers, mysteriously vanished off Florida, a response equal to the bigger-than-life events of that banner year was launched: the greatest search and rescue operation ever conducted. Yet no trace was found. The year 1945 has gone down in history as when "Whirl was king." If so, Flight 19 disappeared in a year worthy of its mystery; in a year worthy of the controversy it would engender over the decades to come.

The unlikely story of Flight 19 began on that December 5 as hundreds of wartime recruits were awaiting their honorable discharge back into civilian life before the holidays. This beginning is one tinted with age and framed by yellowed corners, photo stock mellowed by sepia. The images have become blurry, images of 14 faces caught at a cheery and unsuspecting moment.

Hundreds, even thousands, were going about their daily routines at Naval Air Station Fort Lauderdale, Florida. This base was built merely to accommodate the swelled ranks of wartime recruits. Its purpose was to train naval aviators in advanced over-water navigation. As such, Fort Lauderdale NAS was built just a couple of miles from the beach on the northwestern perimeter of the small beachfront town of Fort Lauderdale on Florida's east coast, about 20 miles north of Miami.

By 8 a.m. on any given day except Sunday the base was already buzzing with activity. On this particular Wednesday, a brilliant tropical sun came up to face a warm southwest trade wind. A drone of engine motors resounded over the base as squadrons of dark blue painted Avengers soared east towards the sun's spreading rays. Their destinations varied, but all were heading over a lazily contorting blue ocean. Some would fly off the coast in squadron exercises, others would growl and swoop over mock targets at sea, and still others would hum over the Bahamas' dozen shades of tropical blue water and use the archipelago's hundreds of palm spread islands as landmarks to practice advanced over-water navigation.

At 3:40 p.m. Flights 16 through 27 were still up. One of them, Flight 19, was about 20 minutes behind Flight 18, both flying the same routine triangular course in the heart of the Bahamas. Giant clouds soared overhead. At only 1,000 feet flight altitude Flight 19 was swept by deep shadows and then thrust into late afternoon sunlight. The choppy ocean, agitated by a teasing 35 knot southwest trade wind, blushed and frowned with the passing of each sailing shadow. Sunlight pecked at the Plexiglas windshields and rays unfolded under clouds as if welcoming arms. Grand Bahama should be greeting them just like this now. Breakers should have been visible advancing in squiggly white lines with the surf to crash on her rocks and honey-like beaches. The huge island stood an obtrusive sentinel before the flight's third and final landmark, Great Sale Cay, where they would turn southwest and head straight home. This is what was supposed to be happening anyway right now. But something was wrong, very wrong.

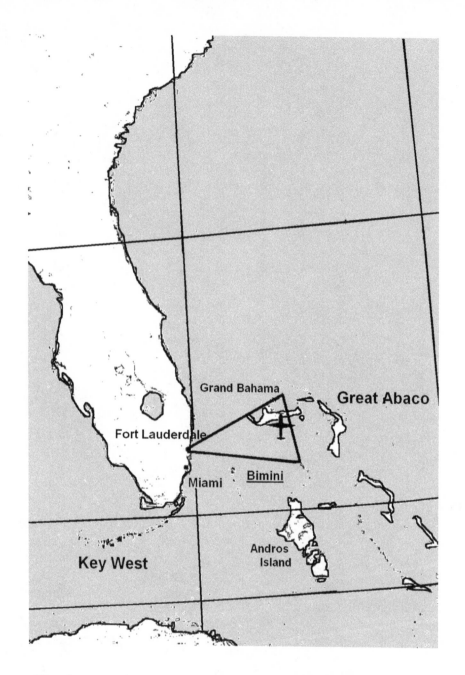

They have not died in vain. . .those whose deaths have inspired a quest.

Chapter 1

Off Florida, Over the Bahamas

December 5, 1945, 3:40 p.m.

PILOT TO PILOT COMMUNICATION CRACKLED FAINTLY OVER LT. Robert F. Cox's earphones. The first voice he intercepted said "Powers, what does your compass read?" The voice asked it a number of times before 'Powers' responded something about his compass. Cox pressed his earphone closer to ear, but the chop of his burly Avenger's propeller was too loud. Soon, however, he faintly heard the first pilot respond again. "I don't know where we are," he said. "We must have got lost after that last turn."

The voices had struggled through airwaves as hazy as the milky tropic sky. Visibility was only about 10 miles. Off Cox's starboard wing Florida's deep green coast was diluted into a pastel watercolor. To seaward a robin-egg blue sky faded into a nylon veil of cobwebs. Against the thin bluish milk a warbird would be as stark as a black eagle. Yet a quick glance over his shoulder revealed only an empty sky.

Few used the training frequency of 4805 kilocycles other than NAS Fort Lauderdale aircraft. Therefore it seemed probable that this was another training flight. Beyond the eastward haze Cox knew other flights were enjoying a kaleidoscope of colors: turquoise seas and verdant Bahamian islands haloed in fleshy beaches. Like him they traveled within their own 10 mile funnel of visibility, cloaked from seeing him as he was from them.

Within his funnel of clear visibility Cox continued to head south to Miami, waiting to hear if the voices would return. The blustery wind resisted him as if an opposing magnet. His Avenger struggled onward, glancing off the invisible gusts before adamantly coming back to course. Great billowing clouds sailed toward Bermuda and never touched the tropic wraith of haze. They sailed on just like the old armadas did so long ago. Their shadows sped over the stampeding Gulf Stream below. The deep blue torrent tried to chase them, spitting up whitecaps and confetti-like streamers in its mad dash at the wind's strident lashing.

Then the voices whispered through the hollow airwaves again. But they did not return to indicate the problem had been solved. Rather the voices returned, joined in a discussion about compasses and headings.

Yet however constricting 10 miles might be to an Avenger traveling the deep sea at 150 mph, most any course in the Bahamas should bring into view a familiar landmark quickly. A pilot— or more than one— not being able to solve his whereabouts by now was cause for concern.

No other voice grasped through the hazy airwaves to bring help. Cox realized he was the only bystander hearing this.

It was a matter of habit that his eyes glanced off his pilot's watch. It was about 3:40 p.m. He grabbed his mike. "Fox Tare seven-four, Fox Tare seven-four to Nan How Able One, Nan How Able One," he called to Fort Lauderdale NAS's Operations Radio, using his squadron handle FT-74. "There seems to be either a boat or plane lost and is calling Powers. Suggest you inform tower of it. Over."[1]

The radio operator at Operations Radio "rogered" rather perfunctorily. There was little need in getting too worried. The operator, too, knew it was practically impossible to get lost for long. If a flight did get lost over the sea, training instructions had engrained it into the pilots to fly west; if over land, fly east and pick up the coast. Simple. If it was a boat it wasn't even their problem. They would simply forward what they knew to Air-Sea-Rescue at Port Everglades a couple miles

---

1 Fox Tare stood for Fort Lauderdale and Torpedo. The number afterward was the squadron number.

away on the coast.

The crackle after the operator's sign-off revealed more faint voices. Cox listened carefully. Now more than two distinct voices could be discerned. Men were discussing headings and compasses. This discussion went on for a few minutes.

"This is Fox Tare seven-four," Cox finally called at a clear moment. "Plane or boat calling 'Powers,' please identify yourself so someone can help you." He released the key on his mike and waited. But there was no response.

According to the clock at Fort Lauderdale's tower it was 4:05 p.m. Petty officer Rolland Koch shifted his watchful study of the incoming and outgoing aircraft. A radio message from a nearby squadron had grabbed his attention. Lt. Willard Stoll, the instructor of Flight 18, was just announcing they were in the landing pattern. Koch referenced a sheet of the ETAs handy. The flight was just a little late. They had taken off at 1:45 p.m. and were due back at 4 p.m. They had flown Problem Navigation 1, about a 2 hour and 15 minute exercise over the Bahamas. The windsock, however, told Koch everything. It was briskly slapping northeast to seaward. The last leg of Problem One was 247 degrees southwest, making the brisk wind on their final course into base a boastful headwind.

Shadows of clouds scampered over the wide flatland field crisscrossed with landing strips to greet Flight 18's arrival head-on. They contorted over the many gray planked buildings of the NAS, as if with animated curiosity inspecting every wall, every crevice, before they sped on. Windows winked with sunlight and then went dark again as another mountainous billow obscured the sun on its playful gallop out to sea.

After radioing Stoll the vector, Koch refocused his eyes in the binoculars and caught a flashing glimmer of light off Plexiglas. The flight of dark warbirds was banking and taking up their position in the queue. Now the "squawk box" rang, interrupting his studious juggling. It was Operations Radio right below him one floor in the tower. "Fox Tare seven-four," began the lazy operator, "has picked up a message from an unidentified source saying they are lost and in trouble."

"What is the identification of this plane or boat?" asked Koch.

"Don't know. Let me ask."

The interruption slightly annoyed Koch. The tower was only concerned with the communication of flights coming and going. Operations monitored all other radio communications from the base. If there was a flight lost, it obviously wasn't in the landing pattern. Koch had the ETA schedule and referenced it. With no more than a mandatory eye he checked what flights might still be further out. There was Flight 17, 19, 20, 21, 22— about 10 more still out, all on various different exercises and courses. At this point there was little reason to get worried. He quickly came back to his immediate duty— monitoring the landing of Flight 18— and left it up to the radio controller on the floor below him.

Operations Radio was a box of a room on the middle floor of the square tower. It was Spartan and bare bones. All the cold gray radio equipment sat upon a heavy wood table against the wall. Some was on the floor and some sitting on other pieces of equipment. There was nothing but box-like windows. Sunlight angled toward the floor and then surrendered to the gray dimness of the room when a cloud whisked by.

The operator put his earphones back in place and slid the mike forward on the wood table. "Nan How Able One to Fox Tare seven-four," he said while working the black tuning knob. "Tower asks if they have any recognition or identification?"

"Negative. Not as yet known," Cox's voice quickly came over.

Lt. Cox had been officious only because he was still trying to understand what the pilots' problems were. He pressed his left earphone receiver tightly against his ear. Moments later, he heard the same voice he had first heard. It came in louder than the chatter had come in between the others.

He keyed his mike. "This is Fox Tare seven-four, plane or boat calling 'Powers' please identify yourself so somebody can help you."

No response. Moments later a voice breezed past the airwaves: "Does anyone have any suggestions?" Cox was sure it was the first voice he had heard. Then the voice said: "I think we must be west of the Keys." [2]

The Keys were hell and gone from his position. The faintness of the voices was now explained. Urgency touched Cox's tone. "This is Fox Tare seven-four calling lost plane or boats. Please identify yourself? Over."

There was that terribly tense moment as the rush of obnoxious static conveyed nothing but desolate emptiness.

Then a voice came over, much louder and clearer.

"Roger, this is Mike Tare twenty-eight."

"Mike Tare two-eight," replied Cox, "this is Fox Tare seven-four, what is your trouble? Over."

"Both my compasses are out," replied the pilot of MT-28, "and I am trying to find Fort Lauderdale, Florida. I am over land, but it's broken. I'm sure I'm in the Keys, but I don't know how far down and I don't know how to get to Fort Lauderdale."

"Mike Tare two-eight, this is Fox Tare seven-four. Put the sun on your port wing if you are in the Keys and fly up the coast until you get to Miami, then Fort Lauderdale is 20 miles further, your first port after Miami. The air station is directly on your left from the port. What is your present altitude? I will fly south and meet you."

There was a moment of that vacuous static and then: "I know where I am at now," said the pilot of MT-28. "I'm at twenty-three hundred feet. Don't come after me."

Uncertain, Cox shook his head. "Mike Tare two-eight," he responded, "I'm coming up to meet you anyhow." Cox waited a moment. Then: "Nan How Able One, this is Fox Tare seven-four," he called Operations Radio. "Call of pilot in trouble is Mike Tare two eight. Over."

Cox unkeyed his mike and thought for a moment. There was reason to be concerned. Just those two words "Mike Tare" had said a lot. "Mike" stood for Miami; "Tare" for a torpedo bomber. A Miami based plane flying over the Keys would not be looking for Fort Lauderdale. Also, any pilot flying over the Keys should easily be able to see the Overseas Highway. It stretches like a long concrete wire, connecting the scattered keys and islands like

---

2 Underlined indicates reconstruction based on radio log abbreviation of dialogue.

beads on a necklace. It was a marvel of engineering which took away the isolated and lonely appearance of the long chain of islands snaking far out into the abyssal blue Florida Straits. Given this, Cox wondered if MT-28's flight was actually over the Bahamas. Any pilot that hadn't compensated for the wind would have been blown northeast, not southwest.

Nevertheless, he pulled the stick and his big Avenger banked and swooped toward the Keys. Heading south against the wind his own hunch was reconfirmed by the surging ripples of the Gulf Stream. They rolled underneath him like a conveyor belt, spitting foam from the strong southwest crosswind kicking them amidships. He felt his own heavy Avenger labor and groan as it roared into the strong headwind.

Faint pilot chatter continued, but strangely it wasn't getting louder. It sounded like the pilot of MT-28 was explaining his problem to somebody else. At one point Cox heard the pilot of MT-28 say that his flight had headed out from Fort Lauderdale on a course of 120 degrees. This, too, reconfirmed Cox's suspicions. On that course, with the prevailing winds, the flight would have been blown northeast over the Bahamas. Cox was beginning to believe he was headed in the wrong direction. His intense concentration on the faint dialogue was then interrupted by Fort Lauderdale's monotone radio operator.

"Fox Tare seventy-four, this is Nan How Able One. Is the call sign of your contact Mike Tare twenty-eight or Fox Tare twenty-eight?"

It was as if Fort Lauderdale was reading his mind. He told them he would find out. He quickly called MT-28. The pilot's response was a quick "Roger, that's Fox Tare two-eight."

With that correction Rolland Koch's suspicion was also confirmed. He knew exactly which flight was lost. It was number 19, commanded by Charles C. Taylor, a veteran instructor. He was on the same navigation problem as Flight 18, which was just now taxiing to a stop by the hangar. This flight plan was entirely over the Bahamas east of Florida and there was no way that the flight could have ended up over the Keys, south of Florida. He confirmed this to the radio operator in Operations.

The crash bell rang in Stations Operations on the ground floor of the tower. Both the Operations Officer, Lt. Commander Charles Kenyon, and Lt. Samuel Hines, the Tower Duty Officer, were jolted in their seats. At that moment the "squawk box" also rang on Hines' desk. It was routine to look at the clock when this happened. It was 4:06 p.m.* Lt. Hines quickly picked it up. It was Operations Radio one floor above them.

"Lt. Hines, please come up to Operations Radio," said the operator. Some monotone routine was now gone from his voice. "We believe we have some planes lost."

Hines went upstairs and asked for the whole story. The operator told him as much as they knew, clarifying that they were not in touch with the planes themselves but that everything was being relayed through the pilot of FT-74.

Hines got on the phone to Nan How Able Three (the Air-Sea-Rescue boat facility at Port Everglades) to advise them there may be a need for them to help coordinate rescue operations. As a part of the Air-Sea Rescue network the Navy and Coast Guard operated, it was a major link with the command station at Miami and was in turn linked with all other stations along the coast. Within minutes, if necessary, it was capable of alerting and marshaling statewide search resources. However, Hines found that the line was busy. He immediately went down stairs and saw that Lt. Commander Kenyon was already on the line with them. He was telling Port Everglades to attempt to contact the flight if possible, as they were having too much difficulty.

Hines therefore went back up to Operations Radio. Just as he entered the room, C.C. Taylor's voice crackled over the receiver. It was calm and professional and his Texas drawl was clear. "Can you have Miami or someone turn on their radar gear and pick us up? We don't seem to be getting far. We were out on a navigational hop and on the second leg I thought they were going wrong so I took over and was flying them back to

---

* The reader will note many discrepancies in time. The clocks were not synchronized, thus giving the impression more or less time elapsed between certain points.

the right position, but I'm sure now that neither one of my compasses are working."

Hines immediately got on the phone to the radar shack. While he waited for a response, Cox's voice came over the receiver in the background. "Fox Tare two-eight, you can't expect to get here in ten minutes. You have a 30 to 35 knot head or crosswind. Turn on your emergency IFF gear, or do you have it on?"

"Negative," replied Taylor.

"Nan How Able One, this is Fox Tare seven-four," Cox's voice crackled through the receiver again. "Flight of five planes' leader is Fox Tare two-eight. He has his emergency IFF equipment on. Requests if he can be picked up on Fort Lauderdale radar gear."

The radio operator's eyes questioned Hines. But Hines still waited. Finally, his impatient scowl disappeared as he shook his head. The voice on the other end had just confirmed their radar was only for in-close training range.

The radio operator keyed his mike. "Fox Tare seven-four, Nan How Able One. Negative. He cannot be picked up on Fort Lauderdale radar gear."

"Roger. Standby," replied Cox. There was that understanding in his voice that this was now getting complicated.

"Fox Tare two-eight, this is Fox Tare seven-four. Turn on your ZBX."

In lieu of radar, this was the only advice Cox could give. The ZBX was the homing device in the Avengers. It was set to a frequency being transmitted by NAS Fort Lauderdale's tower YG (Yagi) antenna. This gear allowed the flight to follow the strength of the beacon and thus come back to base. Depending on the altitude of the flight, the beam could be detected more than 100 miles from base. Radar had a limited range. IFF— Identification Friend or Foe— was but a signal identifying them as friendly, but the ZBX was still the surest way back to the home base.

Yet nothing in response to Cox's suggestion came over. There was only that aggravating static. At this point, he had been heading south now for perhaps 10 minutes and, presumably, Flight 19 had the sun on their port wing and was following the islands in his direction. Yet there was nothing but that lousy void of static. "Fox Tare two-eight, do you read?" Cox called out again. "Turn on your ZBX." Still nothing.

It was now 4:23 p.m. So far, Fort Lauderdale had had no contact with the planes; all they had was a logbook wherein the operator had hastily jotted that the Flight leader is lost and "believes he is west of the Keys."

Lt. Hines decided to get on the ATC (Air Transport Command) line with Boca Raton AAB up the coast. They had working radar that might be able to pick up the flight. "They are on 4805 kilocycles," he advised. "The leader's call sign is FT-28. They were on a navigational hop east of Florida, over the Bahamas, when they got lost." All Boca Raton could promise was that they would try and get their radar shop up and running.[3]

After a frustrated sigh, Hines thought it best to have the radio operator tell Taylor to have a pilot with a good compass take over the lead.

A perfunctory "Roger" came back over from C.C. Taylor.

★   ★   ★

Cox was now south of Miami. His communication with FT-28 was only getting worse instead of better. He heard Fort Lauderdale tell Taylor to have a pilot with a good compass take over the lead, but he did not hear Taylor respond. He waited a moment and then repeated the instruction. Still no response.

With everything that had now happened, Cox was certain FT-28 and his flight was nowhere near the Keys. He was going to call again. However, Port Everglades was now cutting in.[4]

"Nan How Able Three to Fox Tare two-eight," hummed a crisp voice through Cox's earphones. "This is the time for all good men to come to the aid of their country,"[5] recited the operator in concise albeit automated tone. "Can you read us?"

"Affirmative," responded Taylor. "We have just passed over a small island. We have no other land in sight. My compass is

---

3 Many of the long range radar sets had been dismantled after the war, leaving only training range apparati.

4 It was 4:25 p.m. precisely by their clock.

5 Standard line spoken in raising a contact.

out. I believe I am in the Keys. Visibility is 10 to 12 miles." Moments later. "I am at angles 4.5. Have on Emergency IFF. Does anybody in the area have a radar screen that could pick us up?"

Despite his earphones held closely to ear, Cox only heard the part about radar. He keyed his mike. "Fox Tare two-eight, your transmissions are fading. Something is wrong. What is your altitude?"

"I am at 4,500 feet," repeated Taylor.

At this inopportune moment, Cox's ATC transmitter conked out. He immediately started switching channels to try and raise FT-28, but without success. Without knowing it, Cox had just played a crucial role in what would become the greatest mystery in the annals of aviation.

Chapter 2

Hazardous Duty

Fort Lauderdale NAS, December 5, 8 a.m.

THE UNLIKELY STORY OF FLIGHT 19 BEGAN THAT EARLY
morning. Reveille sounded, piercing the quiet dawn and the
methodical roll and boom of the distant surf. A modest breeze
rustled scratchily through the fronds of palm trees. In such a
skittish wind the US flag unfurled lazily to seaward. Shadows of
indolent billowing clouds drug themselves over the barracks of
the NAS and tentatively sampled the booming curl, rippling in
line after line.

By 8 a.m. scores of pilots in full flight gear had ushered forth
and casually walked along the sandy promenades. Naval ratings,
dressed in their denim pants and shirts, with white hats daringly
cocked back on their teenage locks, were joking in friendly
conversation. Some headed to ground schools; others followed
the course of the wind to the large hangar. Even from this distance its cavernous mouth hummed with plaintive moans.
Churning eddies howled through its steel girders and around
corrugated iron. Outside its huge open bay doors rows of blue
Avengers stood at silent ready inspection. Next door the briefing rooms were packed with squadrons ready for their run. As
soon as the briefing was over, and the pilot signed the checkout
chit, propellers turned and exhaust burped under the careful
study of the ground crews.

The Avenger was everything. It was the centerpiece of all this training, indeed the very reason for Fort Lauderdale's existence. The tarmacs were crowded with rows and rows of them neatly lined up with their wings folded back, the painted white stars and FT gleaming in the tepid morning sun. Some already scurried along the runways to take off. The great Quonset-like hangar was very much like an anthill, sending forth and receiving lines of blue warbirds.

Although dwarfed by the huge hangar, the Avenger was a massive aircraft, the largest, heaviest single engine aircraft used in World War II. It was designed to carry 3 men: the pilot, the dorsal gunner, and the radioman. The pilot was separated from the other two up in the cockpit, and there was no possible communication between he and his men while in flight except over the intercom. The dorsal gunner sat within the "tunnel" of the fuselage, his head and shoulders cramped into the Plexiglas bubble with the machine gun. From here he could guard the Avenger from Japanese fighter planes. The radioman sat almost directly under him, encased in the skinny bowels of the aircraft— the "tunnel"— at his radio equipment.

Such was the morning, warm and breezy, and such was the object of all Fort Lauderdale's monotonous existence. And monotony it was. Feelings of pace-without-purpose came with the end of the war. It seemed utterly purposeless to train thousands of men who would soon be leaving the military. It was utterly purposeless teaching them naval tactics that were antiquated by the elimination of any potential Naval adversary in the world. But you just don't sit on your duffs in the military. Men would continue to be trained in the Avenger, and, until the advent of the much anticipated jets, they would continue to be schooled in tactics soon to be eclipsed by those same jets.

Excitement at discharge, however, took the drollery out of this pace without purpose. A special sense of accomplishment graced the highest to the lowest— they had survived the war! They had survived the war to end all wars! The men could taste civilian life coming. Soon they would be back with their families. No more military muster. No more "Yes, Sir" salute, do your duty.

One such man was Robert Francis Harmon. Very much so,

one might add, did this attitude fit "Bob" Harmon. He was as skinny as a beanpole, feisty, always fun to be around, and was perhaps the archetypical "wartime recruit." He was proud to be in the Navy, but he still maintained all the attitude of the "amateur military," a phenomenon only of wartime— a little lax on discipline, a little more the hobbyist than the profession- al. He wore his hat cocked casually back on his head— *way* back on his head— definitely *not* regulation. But he always quickly corrected it to "regulation" style whenever an officer was around— in other words, cocked forward on his forehead.

Typical wartime recruit. . .except for one fact. He was so eager to get into the Navy that he had enlisted while underage and, worse, he also had enlisted under a false name. Now, of course, he had more problems than the usual wartime recruit awaiting discharge. He had to come clean about his real identi- ty for the sake of veteran pay and other benefits, and this in- volved a lot of hassle with Navy paperwork. His real name was George Francis Devlin, and his real age was not 21 but 18. He was so eager to get into the military that at age 15 he took the identity of a man named Robert Francis Harmon who didn't want to enlist and gladly gave his identity to the cocky Brook- lyn teenager.

As a "snot nose" of 16 years old he had seen some rough du- ty in the Pacific as a TBM gunner based off the carrier USS *En- terprise*. The most harrowing of it all was the long night pa- trols. His pilot, Lt. Bill "Smitty" Smith, was expert in the Avenger and was therefore assigned the incredibly dangerous position of being the advanced eyes of the fleet. Both Harmon and their radioman, Weldon Richman, would learn that the night patrols were the most dangerous flying of the war. There was no defensive advantage in night flying. Harmon's excep- tional skill as a gunner was useless since there was little chance of identifying the enemy and, moreover, little if any chance of forewarning. "It's like flying in a fishbowl of ink," Harmon once commented over the intercom. It became one of his fa- vorite visuals. For 8 hours he watched the black morass as Wel- don sat cooped up in the "tunnel" with little more to do than watch his radio apparatus and concentrate on his worries.

The biggest worry was shared by all three— that "Smitty"

might fall asleep. Just one little snort and jolt awake could mean death from a sudden banking of the aircraft, a sudden dive, sudden confusion or vertigo in which it would be impossible for Smitty to regain control. To keep Smitty awake, Harmon sang through the intercom. "It wasn't a distinguished voice," Weldon Richman recalled, "but it was average. He sang, oh, all the popular tunes of the time."

Smitty's other big worry was a navigational mistake. If he ran out of fuel he would have to ditch. At night this could be fatal since an Avenger, for whatever strange reason, had no landing lights. Ditching at night meant feeling oneself down in this "fishbowl of ink," hoping the altimeter was still accurate, hoping the sea was calm enough and then guessing when to let the stick drop and plow headlong into a wavy abyss. Sheer sweating, blind nightmare.

What the Avenger *was* designed for it did well. It was built to be able to come in close to ships and absorb heavy anti-aircraft fire. It was therefore incredibly sturdy. Unlike other fighters or dive-bombers in the Navy, this sturdiness made it easier and safer to ditch the aircraft. For this reason Avenger crews seldom bailed out. If the Avenger was too shot up to make it back to the carrier, then the pilot ditched the plane at sea.

The routine of deplaning was engrained into the men: they unstrapped themselves, kicked open the side door, retrieved the raft from a compartment on the outside next to it, inflated it, and jumped in. By this time, the pilot was usually walking on the wings and then jumped in. They cut the tethering rope and push off. The Avenger could float for over a minute, hopefully, so it was very often safer to ditch than bail out.

But ditching at night, using only instruments in this "fishbowl of ink," gave one little chance of survival. All three knew that. Death was, in fact, all around them, and there was no escape except to make it back to the carrier.

Smitty was, however, an officer who inspired confidence and devotion. All three had trained together at Fort Lauderdale in 1943-44 before being sent to Squantum Naval Base. It was only then that Smitty and "Rich" finally discovered the truth about their gregarious compatriot. While on leave "Bob" took

Rich to visit his family in Brooklyn. After Rich met the Devlin family and in particular "Bob's" father, George Sr., the Chessar-smiling Harmon felt it time to confess he was George Jr. Harmon asked him to keep the secret, so he wouldn't get thrown out of the Navy.

Rich and Bob were pretty close, even though Rich was 19 years old and Bob was only 16 at best. But Rich had no choice but to eventually tell Smitty. Harmon was slated for disciplinary action and Rich thought it best Smitty should understand. A 16 year-old kid is not going to adjust to Navy life easily. He was a Brooklyn Irishman and was cocky, happy, and as the expression was then, "go-lucky." Regimented Navy life, hard for many recruited civilians during the war, was something that Bob sometimes came up against. Bob also overslept, almost habitually. This sometimes required that Rich literally drag him out of bed in the morning. He had to be pulled up, sat up, and be awakened yet again. It was slow motion for him until he finally was completely awake.

Bill eventually became a bit of a mother hen after he found out, and Rich was like an older brother. Together they managed and made a team that weathered the hazards of the South Pacific war and nature's worst elements. When their tour of duty ended, they returned stateside, happy to be away from that damn night flying. However, when they returned the war was winding down. The Navy felt that gunners were becoming an obsolete position in patrol Avengers. Harmon was detached from Smitty and Rich and ended up back at Fort Lauderdale, grounded and itching to get assigned to flying.

Flying was always considered hazardous duty, and with this went more pay. But because veteran gunners were usually assigned only to the veteran pilots, in peacetime a veteran had little chance of getting posted to a flying position. The trainees had their own gunners and radiomen who were being trained along with them. Since training instructors are relatively few compared to the trainees, Bob Harmon had very little flight time. . .and with this he was getting less pay.

This was not the only way in which things had not gone so good for him after he was detached from Smitty and Rich. Oversleeping had gotten him slated for discipline more than once. The only flight time he pulled was a short stint with a trainee, and that had been a harrowing experience for the 18-year old veteran. It was August of 1945 when he penned the following letter to Smitty.

Aug. 26, 1945
Sunday

Hello Smitty

Or should I address you as "Mr. Smith"? Well, here is your ex-gunner, finally getting around, to writing to you. I'm sorry about the delay but I've been having a tough time, down here, trying to get organized. Ever since I left your "protecting wings," I've had nothing but hard luck, everything is SNAFU'ed. (And I thought, Dilbert had a tough time in the Navy.)

I hope this letter, finds your wife, Rich and yourself in the best of health. There are quite a few things I want to say but it's hard to find the words to say them with but I'll try anyway.

First I want to thank you, for being so lenient, with me, when I gave u enough reasons to send me before the "man", like the time I missed all those hops in B.P., I'm really sorry about that, it wasn't done intentionally. Just ask Rich, how long it took him, to really get me awake.

I know I wasn't the ideal gunner but you'll have to admit, I was loyal and added a little ballast in the rear end, (don't forget my singing either.)

My folks and I want to thank you for bringing me back, so many times and in [one] piece. Rich and I are also convinced that you are the hottest pilot, in the whole "God damned" Navy. (Pardon the expression.)

Wish I had the pleasure, of meeting your wife, from the descriptions, I've been getting from the boys, she's really 4.0

Too bad, we never pulled a liberty, together but we may meet again, in the future and really raise some good-old 48 hell!

I've been sweating out a pilot, whose name is Redmond, he's in class 48, which is leaving in three weeks, for a base in Mich. or San Diego. There is a 30 day leave involved so it's not a bad deal.

I'm working on the beach range, while waiting. Can you imagine me working? Things are getting bad. I sure do miss you guys and have bad feelings, for the guy who decided VTN,[6] should be minus gunners but to the navy we are all just numbers.

Time to say "so long" but the best of luck, to you and take care of Rich, radiomen like him are more than just, hard to find.

> Your Ex-Gunner,
> Bob Harmon
> alias
> George Devlin

The last lines reflect Harmon's fear about being permanently grounded. Gunner positions were being replaced by a radar operator. Even radio operators were being phased out. For a peacetime world, there was little use in a torpedo bomber. Thanks partly to their large fuel load and success as long range patrol aircraft the Avenger was being converted solely for patrol duty. . .and this meant no more Harmon.

Now in December 1945 things looked a bit brighter for the youngster. NAS Miami had just ceased flight operations, and so all the trainee pilots and their instructors had been moved up to Fort Lauderdale. Moreover, a lot of guys from Miami NAS had been discharged. At long last Harmon got permanent duty as a gunner flying in Avengers. This time it was with an instructor, not a student. No "sweating" out a pilot when he's already a veteran.

The pilot was Navy Lieutenant Charles Carroll Taylor. Like Harmon he was also a veteran of the South Pacific. He had weathered the war with Torpedo Squadron 7, based off the carrier *Hancock*. He was a cool, laidback hotshot pilot. He had been awarded a number of campaign medals, including 2 bronze stars for his action in the liberation of the Philippines, the Asiatic Pacific Campaign ribbon with 1 bronze star, and the Air Medal, and among others also the Presidential Unit Citation. He had over 2,000 flight hours' experience, 600 in an Avenger, and over 60 hours in actual combat. He had actually ditched an Avenger three times, once without getting his feet wet.

---

6 US Navy designation for "Night Torpedo Squadron."

Chas, Charlie or, more commonly, C.C. stood out— literally. He was a tall, lanky Texan. It was rare to see a 6'1" naval aviator. Pilots were usually short and lightweight. C.C. was tall but he was also lean, only 158 pounds.

Despite his height, Taylor was a natural aviator. He had flown any number of aircraft since he joined the Navy in April 1941, including patrol floatplanes when he was stationed at Key West with Scouting Squadron 62. He had made some daredevil rescues many pilots would never have attempted. Together this revealed the dichotomy that was his character: outside of an airplane he was laidback, but inside the cockpit he was precipitous in action, brave and gutsy, even maverick.

This was the word going around about Taylor— he was seasoned, but he was not precisely "Navy." He was always lighthearted and easy-going. He displayed charming "southern gentleman's" manners. He wasn't snobby, but he impressed people that he was well-bred.

One example of how he stood out is also an example of his easy-going likable sense of humor. It is preserved in his last official photo. His last class at NAS Miami had just graduated in August 1945 at the same time Harmon was penning his letter to Smitty, telling him about some of his woes. The usual class photo had to be taken. He was posed before an Avenger with his graduating pilots next to him— five on each side. They were posed to create a pyramid. The shortest men were on the outside and incrementally the line of officers got taller until there, in the middle, was the 6'1" lanky Taylor. A big, boyish smile characterized this man even at 29 years of age. Usually those photos were staged quickly. But in Taylor's case they took the time to arrange and rearrange everybody to make the statement about his height.

Bob already had one flight with him on December 2. He seemed like a nice guy. Perhaps things were finally going to go smoothly for him after all. Maybe for the time he had left in the Navy, Harmon had found another Smitty.

Well, today there was no worry. With the laidback pace, Harmon was able to get up leisurely and clean up his quarters before chow. He stuffed his white sailor's hat halfway into the back pocket of his denims, tidied up and made his bed. He wouldn't have to don his flight overalls until this afternoon. He flipped on the radio and with a burst of static one of the pop tunes of the day blasted over at a lively beat. As he began to sing along, he could hear the growls and grumbles of those who had not yet turned out.

The music over, the announcer read the call letters and other news of the day. As Harmon left the room, the forecast came over for broken clouds, 67 degrees, a southwest trade wind— flying weather— in other words, hazardous duty pay.

Charles Carroll Taylor

Chapter 3

South of Cuba, West to Pacific

April 1943 to November 1945

THE PILOTS WERE BUT DANCING BOBBLES IN THE GRAY LIGHT, soaked and pitching in the choppy sea. They were little protected by their yellow inflatable. It added only color and a deceptive appearance of warmth, but it was as sturdy as a flower petal juggled by angry rapids, and from this height they looked like two pitiable bugs clinging to it for dear life. Every wave drenched their khaki uniforms and yellow inflated Mae Wests, but Charles Taylor knew they had heard the roar of his plane pass over. The men had perked up as if called from the grave. Though their facial expressions could not be seen, it was clear joy had come to them. Vitality was in every gesture, and one of them almost bounded up and started jumping up on the raft, stretching his waving arms as high as possible.

It was a particularly ironic moment, for C.C. Taylor was in fact circling his own "skipper," William Burke, the Squadron Commander. It had been a day and night since Burke got lost and led his wingman with him into the drink. They had been on a routine patrol, guarding a convoy south of Cuba from German U-boats as it made its way around to the Florida Straits. But Burke and his wingman never made it back. That sounded bad back at base. Anything could happen to one plane, but two? It sounded like enemy attack. U-boats didn't always skulk below the surface. Caught shelling a helpless old scow, they fought back vehemently with anti-aircraft and had taken down some patrol planes.

These seaplanes weren't the most maneuverable. They had huge pontoons that cut down their dynamic moves. Many times their heaviest payload was depth charges and not bombs, practically useless on the surface. Machine guns were then all that a pilot had if he had to duel it out with a sub's belching AA guns.

Burke knew well the limitations of a seaplane. Thus on the second pass over both he and his wingman's brief vigor faded and they became motionless, reduced to men not believing their eyes. Taylor had grinned at them with almost wicked humor, but now he was banking his seaplane to come into the wind. The engine's growling breath was slowing to a steady purr, and the great bird leveled off over a trough in the angry sea.

Taylor's armed patrol seaplane, like Burke's and his wingman's, was far from a rescue plane. Nevertheless, it seemed by the way that stumbling albatross felt out the wind that C.C. Taylor was actually going to try and land in a storm. Burke couldn't help but shake his head in amazement.

The seaplane's pontoons splashed and bounced up and down in the trough, kicked a wave and whitecap out of the way, and then the humming bird came around and headed to the men, its propeller chopping the spray. Taylor rolled back his cockpit canopy and leaned his arm out. He logistically eyed the raft as he coasted in as close as he could. He smiled yet again. That toothy grin was pearly with satisfaction.

Burke ignored his junior's sluicing humor and helped his wingman paddle. C.C. demonstratively waited. He leaned back and crossed his arms for a quick nap— at least a theatrical example of one. Burke was last in, of course. Before sinking down into the back from all but his embarrassment he saw his junior crane his neck around and wink back at him. C.C.'s head then disappeared and out stuck a thumbs-up on an extended arm. Off they surged forward. It was a choppy and bumpy takeoff, but Taylor made it. The trough closed in behind him as his pontoons lifted and cleared the spray. Taylor gunned that bulky albatross and back up it angled into the gray sky and burped farewell to the tempestuous sea.

It was a daring and triumphant rescue, to be sure. More than that Taylor had asserted himself once again as the squadron's navigation officer. His Texas drawl came over the receiver in back. "Welcome aboard, you all." The embarrassment finally left Burke's face and he chuckled at his friend's dry humor.

Two things could always be said about C.C. Taylor. "He was a hell of a flier" and "he had a hell of a sense of humor." He was also one of the gutsiest pilots in the fleet. His rescue of his superior was just one example. His skill at navigation would be seen again when Burke got lost a second time. Taylor finally found him while still in flight. He came abreast him, wobbled his wings and led him back to base.

This was C.C. Taylor, the pilot. C.C. Taylor on the ground and in school was somebody else. Ironically, he hadn't seemed the ideal student at flight academy at Corpus Christi, Texas. At one point it didn't even appear he would pass the muster. It wasn't incompetence. He just didn't like class work— he was barely more than a "C" student— and he was quite laidback for officer material. But once he got into the cockpit he proved himself a different man. Like his aircraft, he sprouted his own wings that freed him from the restrictions of regimented life.

After graduating he was quickly made an instructor and then when it came time to actually send him to his first post, his appointment as Navigation Officer for Squadron 63 at Key West testified to his skill. He had spent about a year at Key West before finally being detached in late 1943. The seaplane patrols had won the sea lanes back for America. U-boats were no longer a major menace and there was no more need for so many squadrons.

Transference brought him to Martha's Vineyard for training in TBM Avengers. The war in the Pacific required carrier based torpedo bombers, and many pilots were now finding themselves in training for landing on carriers. His gutsiness only blossomed more here. During a night practice-bombing he and another pilot, John Germ, discovered they were both locked in their dive on the same mock target (lighted pots). Germ's nervous voice came over the radio receiver and quivered in Taylor's dark cockpit. Taylor casually grabbed his mike and radioed back: "This should prove interesting."

On another occasion a wingman told him he was losing a lot of fuel out of his wing tank. C.C. radioed back casually that it would stop. When the wingman radioed back a while later that it hadn't, Taylor said nothing. The whole squadron was bug-eyed as Taylor's Avenger suddenly rolled and rolled until he had gotten the fuel leak to stop.

Aerobatics in a bulky Avenger simply were not done, but C.C. Taylor always wanted to check out the stability and capability of the aircraft types he flew. By the time he had done the above maneuver he was well aware of the capabilities and limitations of an Avenger. He didn't mind showing off, but there was a purpose to his actions. His apparently "daredevil" confidence, as typified by his rescue of Burke at sea, was actually more the result of knowledge than guts. But he never let on. He preferred the daredevil/mild-mannered paradox.

Such skill, albeit acquired by unorthodox methods, came in handy for him. On three occasions he had had to ditch an Avenger, a feat no pilot wanted to do once. The first time was on June 14, 1944, in the Atlantic. While on the shakedown cruise of the USS *Hancock*, his squadron's new carrier, C.C. got lost on morning submarine patrol. This was a rare moment indeed. Joe Broadwater, his radioman, had always been confounded by Taylor's strange dead-reckoning ability. On every return flight to the carrier, Broadwater dutifully turned on the radar. Yet never once did Taylor call back to ask their position. They always made on it his dead-reckoning alone.

This morning proved otherwise. Yet their own radar now also proved they were too far out to detect the ship. Taylor radioed the *Hancock* and Radar Control ordered him to climb altitude until they could pick him up. Fortunately, he was within the ship's greater radar range. They gave him the right course and he made it back to the vessel. So low on fuel, unfortunately, that he got into the emergency landing pattern but still couldn't make the deck. He pancaked his sputtering warbird into the cleaving wake of the ship.

After the *Hancock* passed its shakedown it was assigned to Halsey's Third Fleet in the Pacific. In November 1944 while the squadron was making its way back to the carrier, Taylor called the commander and asked him to check his navigation, openly stating he disagreed with the course. The squadron commander told him to pipe down, that he was leading the flight and not Taylor. It was a calculated rebuff. Perhaps he also knew Taylor's manner of navigation was always to rely on his own dead-reckoning sense. To many Navy men that was merely a fine-tuned instinct, but no substitute for equipment.

In this case, it turned out Taylor was right. Fortunately for the entire squadron they came upon another fleet of American ships with a carrier and were able to land and refuel.

Taylor's uncanny mental navigation had turned out to be right again. Equipment was fine, as far as C.C. thought. But it was for helping, not controlling. And indeed Taylor's flight record supported his opinions. Getting lost once was below average, and even then he was able to get back to the ship.

Taylor's concentration while flying was uncanny as well. He flew through heavy flak with the same fearless fatalism as when diving in a near head-on collision with Germ off Martha's Vineyard. The object this time was a Japanese cruiser. It was belching out AA flak at all the incoming Avengers. Black, tortured bursts of smoke shook his Avenger with every explosion. Tracing shells streaked the mottled sky and stung at the metal with a piercing twang. It was the Battle of the Philippines and this was the last big stand of the Japanese Fleet. He pulled the torpedo release and pulled out. It was a toss-up whose torpedo actually sent the vessel down. It was later agreed that another pilot's was responsible. For his help Taylor was awarded the Air Medal; the other pilot the Navy Cross. Taylor had come that close to the Navy's highest honor.

By this time, Taylor was one of the most senior pilots in Torpedo Squadron 7. Many of the new pilots were fresh out of training when they arrived on the *Hancock*. They were getting their baptism of fire while he, more than any other, still seemed that imperturbable paradox.

Soon Taylor was a greater rarity. He had survived ditching twice. The second time was November 25, 1944. The *Hancock* was under heavy attack. The squadron was returning from their attack run and was ordered to hang back and wait. The entire squadron veered off and circled at a distance, throttling back as far as possible in order to conserve fuel. Finally they were given the go-ahead. The squadron sped into landing position, but by the time it was Taylor's turn he had to veer off. There was a problem on the deck. There was little chance he could make another pass so he and John Germ had to ditch. Taylor and Broadwater were a little dazed this time. But Harris, his gunner, already had the life raft out and was in it. There he waited for both of them, knife in hand ready to cut the rope. Taylor was quickly back to himself, declaring with a smile: "Well, we made it again!" He strolled along the wing and literally didn't get his feet wet.

Soon after this incident the squadron was temporally based on Faraulep Atoll about 300 miles south of Guam. While he waited it out there, the Seabees came over and needed somebody to fly to Guam for some supplies. Taylor volunteered. Never losing his sense of humor, at the end of the runway Taylor suddenly throttled back his big Avenger and waved to the by-standing men. One ran over and asked what was up. Taylor asked with a smile "Which way is it?" and was satisfied with "300 miles north" in return. His pearly smile grew, and he took off.

C.C. Taylor's next flight to Guam would not be the same. On December 8 Taylor was detached from Squadron 7 to await transference to fighter pilot training, the cream of the crop of Naval aviation. In the interim he flew with Acorn 36. His duties required various ferrying of supplies and planes and test flights. On January 30, 1945, he was flying a banged up Avenger from Ulithi, with James Nafstad as a passenger. The Avenger was basically only good now for training. Possibly not even that.

In any case, Taylor couldn't find Guam. When repeated radio calls went unanswered, he realized his radio wasn't working either. Taylor followed standard procedure and climbed altitude. Although they never got a radio response, apparently Guam heard him or radar had picked him up. Just as Taylor thought they would have to ditch, a PBY found them and informed them how far off they were. It turned out they were about 60 miles from Guam. With a PBY standing by, Taylor now felt a little better about ditching. The sea was very rough with large blue lazy swells. He coasted in as slow as he could and finally hit a large swell.

From the point of view of the men in the PBY it looked as if the Avenger went straight down with no survivors. But it was standard practice to search 3 days no matter what. Therefore, to the surprise of the Navy, Taylor and Nafstad were found. Amazingly, Taylor had pulled off another ditching, this one in the most improbable circumstances. They had spent 25 hours at sea before finally being picked up. Taylor could kick himself for his toothy, facetious smile to Burke two years ago. He was in a worse situation now. As the line was thrown to them from the destroyer, a photographer snapped their picture. "Swell." Taylor, embarrassed, just slouched on the raft and waited for it to be pulled in.

By April something was in the wind. Germany was almost total-ly defeated and it seemed Japan would be next. Fighter pilots would not be needed in such numbers, but it was thought that TBMs would still be useful. Taylor was therefore transferred to Mi-ami as an instructor on Avengers.

Like NAS Fort Lauderdale, NAS Miami was built especially to accommodate the swelled ranks of wartime recruits. The excite-ment of wartime flying was gone, but training pilots was at least something to do. Taylor had remained unchanged by the war. There still wasn't much "Navy" about him. He was probably too easy-going to ever be a qualified ranking officer, but he de-cided to remain in the Service after the war ended. He really had nothing else to do. He decided it with a shrug, knowing he never was better than a "C" student at school.

When in November 1945 Miami NAS was closed and pre-paring to be turned in a civilian airport, the writing was on the wall. His squadron 79M was transferred to Fort Lauderdale to finish training. But it was no secret this base, too, was soon to close. From there Taylor would go where he was sent. Hope-fully, fighter pilot training might come his way again, perhaps even in those new sleek jets that were slated to be the next best thing. Until then he would schlep himself through the boring and anticlimactic duty of routine training. Like all others he knew the Avenger was obsolete and would soon be relegated to the backwaters of patrol duty. Despite his laidback attitude, that regimen was not for the gutsy pilot within him.

Chapter 4

Standard Procedure

Late afternoon, Fort Lauderdale, December 5, 1945

"HE IS NOW ON A NEW HEADING." COX'S FRUSTRATED VOICE, intoned with the hollowness of the radio waves, did little to disturb the routine within Operations Radio. After it crackled over there was static, then: "He's at angles 4.5 and climbing."

C.C. Taylor was following the same standard procedure he had followed during the war. At 4,500 feet he was trying to make himself and his squadron clear on the nearest radar. At this altitude his homing gear also had a much increased range so it could detect the base's beacon. It was standard procedure, of course, and not much else could be made of it. If anyone didn't pay too much attention in Operations Radio it was simply because Cox's statement contained that which was expected.

Right now the problem was on land. Nobody at Fort Lauderdale knew whose radar was working and which radar might be the closest to the flight's position. This was the furthest thing from the crisp efficient routine aboard the fleet.

The only help Port Everglades was able to give was to blandly state another standard procedure:

"Nan How Able Three to Fox Tare two-eight." Port Everglades' operator's voice was intercepted clearly over Operations Radio's receiver. "Suggest you have another plane in your flight with a good compass take over the lead and guide you back over the mainland."

In contrast to the sharpness of Port Everglades' operator's voice, Taylor's was faint and obscured by static. It was merely a perfunctory "Roger."[7]

If Taylor's response was a bit deadpan, it was only because it seemed at the present juncture that a working compass wasn't an issue. The pilots had been debating amongst themselves where they were and thus unsure where they should go.

At a crucial time like this the faintness of the voices proved annoyingly frustrating. One thing mediated worry at Fort Lauderdale: the flight had plenty of fuel and no doubt the problem would soon be solved. There indeed seemed to be little reason to worry. Just fly west if lost over the sea or fly east if lost over land.

This was the immediate opinion of Lt Commander Don Poole, the Flight Officer. He had just arrived at Operations Radio at 4:30 p.m. Kenyon and Hines quickly briefed him, with Hines qualifying that Taylor thought he was lost over the Keys. Based on the proscribed courses of Prob. Nav. 1, Poole instantly dismissed that and deduced that the flight had to be lost east of Florida, over the Bahamas. Standard procedure, as every incoming pilot is briefed, is to climb altitude by 1000 feet for every 10 miles they think they are from base. From the sounds of it Taylor was doing just that.

Then a student's voice came over the radio: "If we would just head west we would get home." That was standard procedure as well. Every incoming pilot was also briefed in this. Come hell or high water, fly west. Poole held up his hand to pause everybody, anticipating there was more to come.

Taylor's voice now crackled over the radio, talking back and forth with Port Everglades. "Fox Tare twenty-eight to Nan How Able Three, one of the planes in the flight thinks if we went 270 we could hit land."

"Roger," responded Port Everglades.

With that being said, Poole didn't see any need to worry. The flight should make land soon enough. He instructed Operations Radio to broadcast to the flight to "fly west, fly toward

7 Spoken at 4.28 p.m.

the sun."

Poole could be brisk, perhaps in this case too dismissive. He wasn't a man who liked to see complications. It may have been too early to be worried about a lost flight, but the fact Taylor had thought himself west of the Keys at one point should have been grounds for concern.

Cox was different. Speeding back to base, he couldn't shake Taylor's early statement. At the moment his transmitter had cut out he had been about 45 miles south of Lauderdale and yet Taylor's voice had only been growing fainter. He was now more certain than ever that Flight 19 had to be lost in the northern area of the Bahamas.

Meanwhile Lt. Azariah Thompson, the Duty Officer at Port Everglades, had also come to the same conclusion. He quickly called Hines. "Inasmuch as FT-74 has run out of communication with FT-28 by proceeding south, I think that this flight is lost somewhere over the Bahama Bank and suggest that the Lauderdale Ready Plane be dispatched guarding 4805 kilocycles on a course of about 075 degrees and try to establish communication with FT-28. And if the Ready Plane can pick up FT-28 better as he proceeds on this course, we will be sure that the flight is lost over the Bahamas. The Ready Plane could also act as a relay on this frequency as it is becoming very difficult to pick up FT-28."

Hines immediately agreed. "I'll dispatch the Ready Plane." The fact Flight 19 hadn't detected their homing beacon was puzzling. Hines added: "Could you ask FT-28 if he has a standard Yagi disk aboard to home in on the Tower's direction finder?"

"Roger."

Hines' concern was heightened by the radio operator's present actions. Taylor's voice was crackling over the receiver, but it was so broken and faint the operator had to tune the knobs to try and pick any of it up. Whines and blurbs issued forth and then there was that monotone sizzle from a jeering blackout.

Hines turned to Poole and relayed Thompson's suggestion. It, too, was a standard procedure for lost plane recovery. However, he was suddenly surprised by Poole's reaction. To Poole everything seemed OK. The flight seemed to be following standard procedure. There was no need to worry yet. Poole told Hines to wait and not send out the Ready Plane.

More than one pair of eyes looked askance at the decision.

There was little time to dwell on it. Port Everglades' operator's voice soon came over the radio. He was dutifully relaying Hines' question to Taylor about whether he had a standard Yagi disk for homing purposes.

Then there was something surprising: "Fox Tare two-eight to Nan How Able Three. We are heading 030 for 45 minutes, then we will fly north to make sure we are not over the Gulf of Mexico."

That was northeast!

Thompson recoiled with surprise. This was obviously not a typically lost flight. He got on the command line with Dinner Key, Miami, a main relay in the high frequency direction finding Net along the coast, and asked them if they had a bearing on the flight yet. Dinner Key's staff might have been a little irked since Thompson had set them to their task just minutes before. Their reply was a quick "Negative;" but they did ask Thompson to "Advise him to send continuously on 4805 kilocycles. We can't pick up his IFF."

Thompson waited a moment to send this out to the flight. Presently, Flight 19's pilots were talking to each other. One asked "Should I drop the last of my bombs now?" to which Taylor replied "By all means." At 4:46 p.m. Thompson finally sent it out. Then three minutes later he was back on the phone with Dinner key. The answer was still negative. Dinner Key now advised that Thompson alert the entire Direction Finding Net.

Even before "Wilco" was out of Thompson's mouth, NAS Fort Lauderdale came over the radio. "Nan How Able One to Fox Tare two-eight, please turn on your ZBX." The voice grew firm and concise. "Repeat, turn on ZBX. Over." [8]

Thompson could only assume that Fort Lauderdale's Operations Radio had also picked up the same alarming message that the flight was heading out 030 degrees northeast. There was a moment of static, and no reply was heard. "Nan How Able

8 The homing gear.

One to Nan How Able Three," then said the Fort Lauderdale operator, "if possible advise FT-28 to turn on his ZBX."

Thompson had the operator send the broadcast. He spoke each word clear and concise. "Nan How Able Three to Fox Tare two-eight, please turn on your ZBX. Repeat, please turn on ZBX." Static. Minutes later. "Nan How Able Three to any Fox Tare in flight with Fox Tare twenty-eight, turn on your ZBX. Over." Nothing. Minutes passed. It was 5:p.m.

Despite Poole's premature assurance 30 minutes before, Flight 19 was in no way heading west when Taylor had radioed Port Everglades that one of the pilots believed they *should* be heading west. If the men at Fort Lauderdale had listened carefully at the time they should have been troubled. Flight 19 had 4 other pilots and each plane had 2 compasses. Why did only one pilot think they should head west? Why were 4 pilots not insisting they should head west? Taylor would not have questioned 4 pilots if they all said the same thing with certainty. None of the pilots were rank rookies. Each had, in his own way, found himself at Fort Lauderdale after a long and sometimes difficult path through the war. Like many spokes connecting to the same hub, each found himself at the same place in time. Together they should have remedied this situation immediately. But now each found himself locked into an improbable circumstance.

George William Stivers

Chapter 5

A Place in Time

Early Morning, December 5, 1945

IT WAS OBVIOUS AT THE JBOQ— JUNIOR BACHELOR OFFI-
cers' Quarters— to any who bothered to look out their windows that it was ideal flying weather. The distant hum of
Avengers was mixed with the sounds of jeeps starting below
and pulling out to head down the long road to the hangar. A
more gentle addition to the noise was the rustling sound of the
breeze in the palm fronds. They were planted just outside in
the sweeping fields of lawn, giving the JBOQ a rather grand
old fashioned promenade. The scent of cut grass wafted in, a
subtle touch mixed with the pungency of oil and gas fumes.

Those officers who knew they weren't scheduled to fly until
the afternoon had stayed up later than the others at last night's
Officer's Club party. The O-Club parties were the only real
social event on the base that united the rotating classes of training officers and their trainees. At last night's party, the NAS
Fort Lauderdale officer's wives got to see the new pilot trainees
and instructors from Miami. Among Charles Taylor there was
the very distinguished Marine Captains George Stivers and Edward Powers, both in C.C. Taylor's Training Squadron 79M
out of Miami.

At 8:30 a.m. the rush in the JBOQ's long hallways was over.
Those remaining were still sacked out or in a slow routine,

schlepping themselves down the hall to the bathrooms and showers.

George Stivers was one of them. He was always popular at the parties and one of the last to leave. He had handsome, noble features, and a captivating way of behaving. He was an "All American" from Missouri— excellent grades at high school and excellent performance in sports. He had always intended to be a professional Marine. Out of all of the students, he was one of the most distinguished, being a graduate of no less than Annapolis Naval Academy. As a freshly graduated 2nd Lieutenant, he had to immediately set aside the Annapolis cadet's dress uniform and hat of a front line soldier from the War of 1812, and in its place he had to put on the camos of a WWII Marine platoon leader in the 3rd Raider Battalion. He was quickly shipped to the South Pacific, where he further distinguished himself by being cited for gallantry— twice on Guadalcanal and once on Tarawa.

"The Canal," as Guadalcanal would be called, was a stinky, malaria-ridden island, and the battle for it was a bloody battle of attrition for supremacy, both on the island and at sea. Then, after this, none thought it possible but the horrors of the Canal were eclipsed by "bloody Tarawa," an island in the Gilberts chain, a key target on America's island hopping campaign to Japan.

Hitting the beaches was a nightmare at Tarawa. The landing craft were stopped by bulwarks the Japanese had entrenched at the surf's edge, preventing them from driving up the beach. Marines had to leap over the sides of the craft and were gunned down in mid-air before they could take cover. Men were hit, calling out for their mothers. While taking cover, surrounded by ricocheting bullets, one might see a soldier trying to stuff his intestines back into his abdomen, before he slumped over. "Corpsman" was being shouted constantly. Bullets burned and streaked to their targets. Entire platoons were cut to shreds trying to destroy pillboxes. Many a Marine never spoke of Tarawa or the Canal. . .and many of them hit the bottle pretty hard to forget what they had seen and what they had had to do.

Stivers was no exception. Memory becomes a curse for those trying to cope with the after-effects of war. It is a bizarre reality to see thousands of human beings trying desperately to kill each other. You know that the other guy must be thinking what you're thinking— "When is it my turn?"—and then he suddenly falls down dead next to you. How insignificant you feel. You're only a target. Then to know that your enemy must be thinking the same thing. Then, in your sights, you kill your first man. You know full well that you just stopped his own thoughts of "When is it my turn?"

It was stuff like this that Stivers still had to contend with as an aviation cadet. He was no alcoholic, of course, but he hit the bottle hard at the parties, sometimes too hard. This led, in one instance, on November 5, 1944, while already a student aviator at Pensacola NAS, Florida, to discipline for being "intoxicated in public." He was confined to quarters.

The words of the Letters of Commendation for Gallantry are of little comfort for those like Stivers who truly earned them. Washing his face in the bathroom, preparing to shave for the day's flight, he could see in the mirror a much-aged version of himself; no longer looking 25 years old, but rather closer to 40. The vaunted words of his citations echoing distantly in the vault of what we like to poetically (or euphemistically) call the recesses of the mind called him momentarily back to the events that had so aged him.

For gallantry in action on Guadalcanal, British Solomon Islands. During the period 2 January to 4 January 1943, you volunteered to repeatedly lead a small patrol beyond your battalion's defensive position and assisted materially in mopping up the remains of an enemy force and in capturing considerable quantities of Japanese arms and ammunition. The courage and determination displayed by you during the operation were in keeping with the best traditions of the Naval Service.

Again he would be commended for gallantry in December 1943 for "meritorious conduct during action against Japanese forces on Tarawa, Gilbert Islands, 20 November 1943." The formidable commander General Julian C. Smith would personally sign the written certificate.

Luckily Stivers was able to become the Aide to the Assistant

Division Commander in January 1944 and get away from the bloody front lines. It was better for him. It was becoming apparent that he was suffering from ill health. In between such action and gallantry and the resultant praise, Stivers had been taken ill several times from malaria. Even after being detached from front line duty, illness dogged him. In April 1944 he was promoted to Captain. . .while sick yet again in the dispensary. The post dispensary was not enough when in July he became severely ill again. He was sent to the Naval Hospital at Norman, Oklahoma. Again, in August, he was sick at the post dispensary.

In October things finally started going good for him. He was able to make the change to naval aviation. He would become a Marine pilot in training at NAS Pensacola, Florida. This way he could remain in the military and still do his duty in the war. If he had to fight the war, it was better to do it from the distance of an airplane. Your target is merely that— a target. There is no seeing the man close up; there is no smelling your enemy and killing him hand-to-hand; there is, most of all, no seeing your buddy blown away before your eyes.

Yet throughout all his illnesses and trauma his personality was never dampened. He was the outgoing Marine captain, the life of the party, and he was always gregarious and energetic. He had more flying time in his training squadron than any other pilot. Sickness, booze, partying— nothing kept him from his job. A rare man was Stivers. Now that the war was over, he could really breathe a sigh of relief. Bloody World War II was over thanks to the atomic bomb. His health was coming back to him, and the future seemed a peaceful one. Just finish today's mission. He was now, in December, a very relieved Marine.

Jimmy Gerber was also still there at the JBOQ. He is another example of the veteran turned aviator. He had joined the Marines just after Pearl Harbor, and was then officially inducted in January 1942. After being trained in San Diego, he was stationed stateside at Bremerton, Washington. His native Minnesota background made him an ideal choice for cold weather combat training and thus he was soon transferred to Adak, Alaska, as a part of the operation to keep the Japanese from taking any other islands in the Aleutians chain. He was a very capable Marine and also was well liked. He made it only to corporal when, in 1943, he got recommended for officer training in Naval aviation.

Attaining naval cadet status from corporal is quite an accomplishment for anybody, especially since Gerber's career during the war had not been remarkable. He was, however, a crack marksman. Gerber's acceptance as a Naval cadet was a reflection of his keen eyesight, coordination, and his leadership abilities, something that was quite evident to his commanding officers at his various bases.

Gerber's training finally ended at NAS Corpus Christi in July 1945. He was now a fully trained pilot in the United States Marine Corp. With this came the promotion to officer. He was immediately transferred to NAS Miami, there to join Squadron 79. He was now where he wanted to be— far better than being a mud Marine on one of the islands. If he only felt this before joining, he knew it for a certainty now after a rare moment when Stivers might uncork a short story about Tarawa or the Canal, before choking up and abruptly stopping.

The "snot nose" of the whole bunch was Ensign Joseph Tipton Bossi. He was the genuine farm boy from Kansas eager to dust the hay off and get into Naval aviation. On December 14, 1942, just shy of his 18th birthday he had left the University of Kansas and enlisted in the United States Navy. Although he had barely any college education, his skill at flying, his personal character, and his enthusiasm clearly marked him for officer and pilot training. This, in turn, led to him being qualified for Avengers.

Bossi had taken to flying like a natural and to the Avenger in particular. Just two days before, on December 3, 1945, he had written to his parents letting them know that he passed up his chance at a discharge. "I would have no chance to fly planes like these if I came back to Arkansas City," he explained to them. He loved everything about the Avenger especially when compared to the dull routine of Kansas farm living. When volunteers were called for to ferry planes to Arkansas he was one of the first to raise his hand. He had written his parents that until after Christmas there should be no more of those long flights, but he really didn't mind them.

Today's flight certainly didn't fit that category. It would be only around 2 hours and 15 minutes. While the others were still groggy Bossi was already standing before the roster board to find the precise flight time: 2 p.m. That meant that they would be back by 4:30 p.m., graded on their plotting boards (used when navigating), and then by dark (5 o'clock) be back at the JBOQ and changed, off to dinner and a movie across the street.

Today's hop— Problem Navigation 1— was, in fact, very easy. It would be the third and last of basically similar maneuvers. Each was a triangle over the Bahamas comprised of "legs"— distances between each landmark. Each student led for a leg and then turned over the lead to the next at the following landmark. If someone missed their landmark, they were all in big trouble. The entire group would get a "down" score because it meant the other pilots hadn't been paying attention and following the courses flown by the leader for that leg.

Prob. Nav. 1 required that they depart Fort Lauderdale at 91 degrees True. Head out to sea for 56 miles, where, northeast of Bimini, they would conduct practice dive-bombing on an old shipwreck. Then they would proceed further easterly for 67 miles. At Great Stirrup Cay turn to a northerly direction of 346 degrees for 73 miles, cross Grand Bahama and then within sight of Great Sale Cay turn southwest to 241 degrees and fly that heading for 120 miles right into Fort Lauderdale. At one point along this last leg they would cross Grand Bahama again at its tip, about 65 miles from the US coast. Easy.

Aviation Radio Man 3rd class Walter Reed "Tony" Parpart checked the flight schedule board as well. All he had to see was that Taylor had a scheduled flight to know that he had to work that day. He was Taylor's permanent radio operator. While Harmon was up in the dorsal turret bubble swirling around just

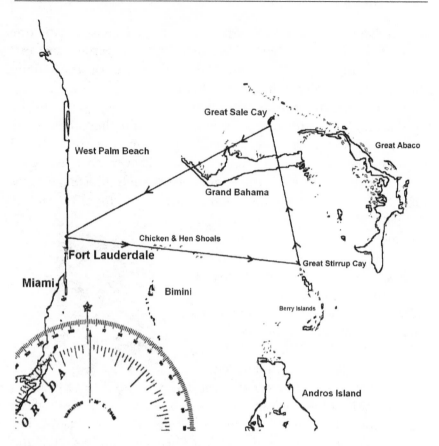

The exact route of Problem Navigation No. 1. Land was visible many times during the route: at Chicken and Hen Shoals, Great Stirrup Cay, thrice at Grand Bahama, and once at Great Sale Cay.

over his head, "Tony" Parpart would just be sitting at his radio equipment or reading a magazine. There was little else to do on a training run. The student radio operators had to undergo certain tests while in flight. But for a seasoned veteran like Parpart it was just a question of riding along and clocking in time and extra pay for hazardous duty. . .unless one of the students had real trouble in their plane trying to home-in on a signal and work the code. He wasn't even needed to contact the other pilots. Taylor had a receiver/transmitter up in the cockpit and used that to contact the others.

Despite the hazardous duty pay, a lot of radiomen were tired of flying. Indeed, for this hop today the radioman wasn't even needed. On previous hops they had to pick up Fort Lauderdale's various homing beacons and then return an appropriate signal. But not this hop. The gunners, like Harmon, would have something to do— at the very least swirl their turret around and sightsee— but today the radioman was only ballast locked in the narrow "tunnel" of the Avenger.

Edward Joseph Powers Jr.

Chapter 6

Born Leader

"ALL PLANES IN THIS FLIGHT JOIN UP IN CLOSE FORMATION."
C.C. Taylor's voice hummed through Ed Powers' earphones.
He looked out his cockpit and saw the big blue warbirds close
in around him. At last, Taylor's swooped in and leveled off behind them like a sheepdog keeping its charges in line.

"How long have we gone now?" Taylor then asked Powers
specifically.

Powers looked at his watch— 5:04 p.m. He could not help
but shake his head. "About 20 minutes," he replied into his
mike.

Powers' voice was tightened by frustration. This all seemed
pointless. For 20 minutes they had been flying northeast, and
he was certain they were going out into the Atlantic. With
each minute that passed a clenched fist was only tightening
more within his stomach. Taylor's hope that this course was
bringing them across the Gulf of Mexico to the Everglades of
southern Florida was absurd. He may have sounded his usual
confident self, but this time he was confident in error.

There was one other thing wearing on Powers' mind. With his
eyes watching the ocean swells skipping along with them, he
couldn't help but think of the Belvito incident. Michael Belvito.
The Belvito Affair.

Powers could not prepare for a flying assignment without thinking of Belvito. Only 2 months ago on October 9, 1945, Powers was one of 6 pilots participating in a bombing drill over the old shipwreck near Bimini. He had only one crew this day, his gunner sergeant Michael Belvito. The drill routine was very specifically laid out in the briefing room before they left. They were to ascend to 4,000 feet, glide bomb over the old wreck, then level off for a "masthead" torpedo run. This would be a mock maneuver of a real torpedo run on an enemy ship. Come in low, glide like a gull over the blue sea, release the torpedo and pull out.

It was Powers' turn. He released his smoke bombs and continued on. Suddenly, he felt a thump, followed by a burst of air whipping into his cockpit from under the narrow space below his seat. This was the only space that connected his cockpit to the rest of the Avenger tunnel beneath and behind him. He hadn't a clue as to what happened, but continued his run, leveling off and speeding straight for the dummy hulk. He "released" his torpedo and climbed back to position.

After the entire squadron regained altitude, he saw that a couple of the other pilots were trying to get his attention both visually and, by demonstratively holding up their microphones, trying to contact him over the radio. A couple were pointing at his fuselage's tail section. Powers couldn't fathom what they wanted, if anything. Nothing was coming over the radio. After landing back at base, he jumped down off the wing to the awaiting group of pilots. They were gathered by his plane's fuselage looking inside. His side door was missing and so was Belvito! He had disappeared!

Of course, there had been an accident investigation of the Belvito incident. "It is the opinion of the Aircraft Accident Board that Belvito voluntarily and precipitately abandoned his aircraft during recovering from the second glide bombing run; it is believed that Belvito, not realizing that the division was changing from glide bombing to masthead bombing, evacuated the plane because he thought his pilot was going to crash into the sea following the second glide bombing run when the aircraft did not climb for altitude but proceeded to get into position for masthead runs."

How could anybody have been that, well, dumb? As a former training instructor, Powers was used to seeing all kinds. There is the old adage of "the Navy 5 percent." No matter what you say, there will always be 5% that never hear you because they don't listen. Accepting that Belvito did not bother to pay attention to the briefing, why did he not get on the intercom and verify what was going on? Was it not working, too? The other pilots could not raise Powers either. Those blasted radio dial knobs had 12-foot cables and over time they get out of adjustment and have too much play. Sometimes you think you are right on the frequency when you aren't. Did Belvito try to but failed to contact him and therefore thought Powers had already abandoned the plane?

At that altitude Belvito did not stand a chance. It is probable his chute didn't even open. A full search had failed to find a trace of him.

Powers was naturally deeply affected by Belvito's loss, especially his loss in such a manner. He was not only the pilot but as the commanding officer of the Marine detachment his superior at all times. In both these capacities, he wrote a touching letter to Belvito's family in New Jersey, condoling with them on their loss. There is no question from the tenor of the letter that Belvito would remain on Powers' mind every time he had to prepare for a flight over the sea.

This was not the moment for Taylor's impatient voice to burst over the receiver.[9]

"Let's turn and fly east two degrees. We are going too damn far north instead of east. If there is anything we wouldn't see it."

Before Powers could even grab his mike there was vying from each pilot to be recognized. Powers finally got a word in. "Dammit, if we would just fly west we would get home!"

Tension mixed with quandary in Operations Radio. For the last 10 minutes Operations was in a state of flux. When Taylor had ordered them northeast Poole had relented and ordered the Ready Plane to prepare. His trigger finger had been tensed to give the order to takeoff, but then Hines had rushed in. "Hold

---

9   5:05 p.m.

it!" he blurted. "Commander Baxter says they will have a radio fix for us." (Baxter was the Assistant Commander of all Air–Sea Rescue HQ at Miami.) Poole then ordered everything to 'standby' and relayed the news to Thompson, who had prepared 3 PBMs at Dinner Key to takeoff. None of them had known where exactly to send rescue planes, but it seemed Baxter would now solve that problem for them. They waited. As soon as the word would come Poole and Thompson would order the planes to takeoff.

. . .But they were still waiting, and now Taylor had ordered the flight direct east! As the minutes ticked by, the tension grew. Where could that flight be? Poole remembered that Boca Raton had a huge antenna. Their equipment could pick up the flight's emergency IFF signal as far as 150 miles out. Poole gave them a call.

Unbeknownst to those at Lauderdale's Operations, Thompson had a similar idea, though one that covered the opposite direction. If Flight 19 was around the Keys or in the Gulf, as Taylor still thought, Boca Chica, based at Key West, likewise could detect them with its huge radar range. This would let them know for sure if the flight was within 150 miles of the Keys. It was a good thought, even if it was a very late inspiration.

As both stations continued to wait for Baxter, it was easy to calculate that Flight 19 had only gone northeast for 22 minutes. They could not be that far from that lone island Taylor had reported earlier. They still had more than enough fuel to get back to the mainland. Follow that one pilot's testy advice. By all mean— Dammit, head west!

Moments passed. Taylor's voice broke over: "Fox Tare two-eight to all planes in flight, change course to 090° for 10 minutes."

That was direct east!

The reaction at Operations Radio would have been more incredible had everyone not been trying to hear the faint inter-squadron chatter of the pilots engaging in some heated discussion.

No matter what it was about *precisely*, in only 4 minutes after heading east, at 5:11 p.m., the upshot of it was clear.

"You didn't get far enough east," protested Taylor. "How long have we been going east?"

C.C. Taylor was being challenged. For those who wish to disagree that Ed Powers took over the flight the actual dialogue does not afford them any room. The gaps between the dialogue are equally significant. Taylor had given orders to fly east for 10 minutes, only 4 minutes afterward Taylor could do nothing more than protest. A Marine Captain and a Navy Lieutenant are equivalent ranks. In such a circumstance, the inevitable challenge emerges: "Date of commission?" Taylor would have to say "April 1, 1944." Stivers would say "April 30, 1944." At that moment Powers could say with some finality "November 1, 1943." Powers *had rank by seniority.* Although controversial because Taylor, as the instructor, was the commanding officer of the exercise, Powers was by seniority the superior officer.

Those who choose to still question whether Powers had the courage to assume command know little of the wheat-haired young man from Mount Vernon, New York. Like George Stivers, Edward J. Powers Jr. was the furthest thing from a rookie. He had enlisted in the Marine Corps in 1941, even before the war, after graduating from Princeton University. His qualities were such that for the entire war he had been a training instructor at Quantico, Virginia. In this capacity he prepared thousands of men for war in the bloody South Pacific.

Yet though he was a fine officer, Powers was an opposite of Stivers. He had never seen any action whatsoever during the war. It was a big step therefore for Powers to request a transfer in 1944 to become a cadet in the naval air forces. He gave up his commanding position to assume the position of a lowly trainee. After 3 years of constant command, this was indeed a big step.

Powers had no bloody images of Tarawa or of the Canal, but he had heard enough from those coming back. Thus he was very well aware of what could happen to him when he requested his transfer. As far as anybody knew the war would last a long time, perhaps until 1947, until the Japanese homeland would be conquered. That would be the bloodiest battle of all. The Japanese were training all the citizenry to fight, even if only issuing them bamboo spears. American pilots who were shot down were not likely to survive.

It would seem foolish to question that Powers asserted his rank, especially in light of the following dialogue.

At 5:12 p.m. C.C. Taylor's voice came over Port Everglades receiver. "Hello Nan How Able Three, this is Fox Tare twenty-eight. Do you read? Over."

"Roger. This is Nan How Able Three. Go ahead."

"I receive you very weak," replied Taylor. "We are now flying two seven zero [West]."

"Roger."

Several minutes passed. There was some dialogue, but none was picked up clearly. It didn't seem it was meant to be. It was inter-squadron chatter again.

Finally, at 5:16 p.m. Taylor's voice came through the receivers at the bases again. "We will fly two-seven-zero degrees until we hit the beach or run out of gas."

It was spoken definite, unwavering. There was no hint of protesting. The argument, which many listeners assumed had happened, was obviously quickly settled.

The ordeal was finally over. The flight should hopefully be back within an hour. Relaxed chatter followed a collective sigh of relief at Operations Radio. Then the basic "How the hell did all this happen to begin with?" became the topic. Poole now felt more justified than ever in holding back the Ready Plane.

This was the wrong moment for Robert Cox to walk in.

He was just coming up from Stations Operations where he had been telling Lt. Hines exactly what he thought had happened to the flight. Cox was sure that if he took the Ready Plane to the northeast he could contact the planes and bring them back. Hines was rather indefinite about it and, in the midst of other routine duties, simply said that "Mr. Poole was in charge of it." After Hines scurried off, Cox went upstairs. It was at this moment he told Poole his hunch. Poole almost blurted "No! . . .I don't think there is any use in sending it out now."

Cox backed away from the rebuff and hung around by the wall, trying to listen. There was other chatter coming over the radio, but nothing from Flight 19. Most of the officers weren't even paying attention to it anymore. The problem was essentially solved. But Cox waited and listened. . .just in case.

Chapter 7

Gallant Hours

LITTLE WAS COMING OVER THE BASE RADIO RECEIVER FROM Flight 19 perhaps because there was nothing to say anymore. The decision to head west was final, and Powers was no doubt backed up by Stivers and perhaps also by Gerber. Taylor's last message truly said it all, didn't it? Unless something untoward happened, there was really no reason for Taylor to inform base and update them on circumstances.

Each pilot was silent and brooding in the aftermath of an event which could get Powers court-martialed and a brave instructor suspended from flying. Each crewman was also probably glad that he was far-removed from this controversy. However, most of them were as equally seasoned as the pilots and thus they were used to far more alarming SNAFUs than getting lost for an hour.

Today's little mishap was certainly nothing to Ed Powers' gunner, sergeant Howell Thompson. Like Stivers, he had a veteran record in the South Pacific. He had entered service on December 9, 1942, and had been an aircraft engineer with Squadron 23, airbase group 25D. From there he saw action with none other than fighting squadron 214— the famous "Black Sheep" Squadron. He was a member of the ground crew of its third incarnation. This was just after Pappy Boyington ceased to command it. One of Boyington's pilots, Stan Bailey, was now in command.

By Thompson's time Black Sheep Squadron was based on the carrier *Franklin*. Few ships or crews suffered as much and still kept their ship afloat. On March 31, 1945, the *Franklin* fell victim to an unexpected attack off the Japanese coast. A bomber dived through cloud cover and dropped two 550 pound bombs that ripped her apart. Gutted and stained by hellish fires, her stern flight deck was a heap of gnarled iron and gashed by huge craters from internal explosions. The final body count was 772 crewmen dead. Among these were 32 members of the storied squadron. Thompson was lucky to be one of the men who survived. Along with all the other crew he had fought desperately to save the ship.

It was because of this attack that Thompson finally decided it was time to get on the shooting end of the war. On July 11, 1945, he was transferred to Miami to become a TBM gunner, but by the time he graduated the war was over. On the 15th of November, when Miami closed, he, along with all the other trainees, was sent to NAS Fort Lauderdale.

Powers' radioman was George Paonessa, a solid, sturdy Marine who had joined up on February 26, 1943. He had been stationed at Jacksonville with Training Squadron 10 as an aviation machinist mate (mechanic), where he remained until February 22, 1944, now with the rank of sergeant and qualified as an airborne radar operator. From there he eventually ended up at NAS San Diego, where he was trained and qualified as a radio gunner on a Douglas Dauntless. With this came action in the Pacific with Squadron 151 at Ellice Islands and then Kwagalein, Gilbert Islands, from July 1944 to May 1945. On August 13, 1945, he was transferred to NAS Miami, where he studied the radioman's position in the Avenger, but he dropped the course because of his inability to learn the code. However, on September 29, while still at NAS Miami, he passed as a gunner in the Avenger.

George didn't like failure, and he loved to fly. Therefore with the war over it was easy enough to retake the courses for radioman. This time he was determined to learn the code. He thus got into Squadron 79M as a radioman trainee again.

Marine Captain George Stivers' crew consisted of an interesting duo. The gunner was Robert Francis Gallivan, another seasoned Marine who had seen action in the South Pacific. He was yet another one of the early recruits inducted into the Marine Corps (on February 21, 1942), after Pearl Harbor. He was born in North Hampton, Massachusetts, and after Boot Camp he received training in TBMs at Naval Training Center Great Lakes and then at Naval Air Station San Diego, California.

Gallivan was a compact, feisty Irishman, with a boyish, freckled face. He wasn't big enough for the usual front line Marine (they were usually around 6 foot). Per usual with the smaller but muscular guys he was placed in some branch of aviation. From December 15, 1942, he was attached to Marine Squadron VMSB-143 as an airplane armorer and torpedo repairman. This landed him at Henderson Field on bloody Guadalcanal in the Solomons, where he saw some of the toughest battles of the Pacific War.

The nucleus of the whole battle for Guadalcanal was always Henderson Field. The strategy was to capture and hold it, the only airport on the island, or make the one's holding it incapable of using it. Control it and control the airspace. Control the airspace, win the battle. Either side was lost without it. Henderson Field initially yielded to the Americans, but this didn't create a sweeping victory. American manpower and naval power was too thinly spread out to swiftly dislodge the Japanese from the island. Their troops continued to maintain a foothold, intermittently being fed and re-armed by the Tokyo Express.

Henderson Field was perhaps the most forward airstrip in the war or any war. It was not some airstrip behind the lines in safety. It was the object of the battle. During the day, the blue sky was crisscrossed with smoking aircraft trails or spirals swirling down to the sea as the planes plummeted out of the convoluted dogfights. Fighters dived in howling whines, their chattering machine guns painting the sky with contrails of bursting smoke. Ground crewmen, like Gallivan, had to run for cover when Japanese fighters came swooping suddenly over. For artillery, there was little warning. Great naval battles echoed off the island at night. Dynamic fireball bursts pinpointed where the guns were unloading iron piercing projectiles into the hulls of other ships. Sometimes Japanese ships got through and pounded the airfield without warning. Shell explosions, dirt and rocks hurling at shrilling speeds, sent men racing yet again for safety.

Then there were the heavy rains; then the dysentery; then the fear of snipers; or worse, the fear of another evacuation as Japanese troops might gain the upper hand and drive the Americans back into the sea.

Well, that was the bloody Canal. It was in this environment that the young Gallivan was busy trying to keep the mass of aircraft armed at Henderson Field. From there he advanced along with the island hopping campaign— next to Bougainville and then to the bloody hell that was Tarawa in the Gilbert Islands.

Starting May 5, 1943, he was hospitalized for an extensive period due to sickness from battle and from disease. However, he was tough. By 20th of September he was out and transferred to duty on VMTB-143, again as an airplane armorer, advancing with the spearhead of the US Navy.

As the war was drawing to a close, he decided to transfer to become a TBM gunner. In this he succeeded and was sent to NAS Cherry Point, North Carolina, for preliminary training and then in August to NAS Miami to finish his courses. In September, before his training was through, the atomic bomb brought the war to a close. He had no desire to remain in the Marine Corp. Like the millions of other GIs, he quickly put in for his discharge and prepared to say goodbye to the war years. It was finally approved in November and the papers came through. He would leave in February 1946 and return home to his dad, John Gallivan, who considered "Bobby," his only son, to be "all he had." This flight, of December 5, 1945, was to be his last hop. It was just a routine puddle jump of 2 hours duration. He was going to call home soon and have his dad "turn down the bed and prepare for his home coming." He was coming home for Christmas!

Gallivan's companion on the training flight was quite a bit different. He was Robert Peter Gruebel from peaceful Long Island. "Yo-Yo," as he was nicknamed because he loved to fly so much, had never seen any action. He entered the Marine Corps on 23 April 1944 when Gallivan was already in the thick of things. He was, however, more used to flying than Gallivan. He had been trained from the onset of his enlistment to be an aviation radioman and radar controller with VTSB-931.

Surprisingly, Gruebel's desired duty was to be an aerial gunner. . .surprising considering he was a priest-in-the-making, a calling that the horrors of war only reconfirmed for him now that he was leaving the service to join the priesthood. Ironic that his love for flying would place him on the roster as a last minute fill-in for Flight 19— not as a gunner which was his current duty but as a radioman. He was glad to volunteer for the vacant seat in Stivers' plane. Any chance to fly!

Same thing happened to Jimmy Gerber. His radioman didn't show or was discharged, and the Aviation Training Officer, Lt. Arthur Curtis, couldn't find anybody to take his place. It wasn't considered a big deal. It didn't affect Gerber's qualifications as a pilot. Gerber would simply fly without his radioman for that hop.

Billy Lightfoot, a thin, stringy kid from Oklahoma, would be his gunner for the day. He had been in the Marines for over two years, and was trained as an aviation mechanic. He knew "Yo-Yo" from Squadron 931. Both had been in Marine Air Group 34 together from March to July 1945.

Like Paonessa, Lightfoot wanted to get into the flying aspect of the TBM and not just remain its mechanic. He graduated from gunner's school at NAS Miami in August 1945 and was now completing his advanced training. It is doubtful he told his mother about his new post in the Marine Corp. Billy had a brother, Eugene, who was a B-17 pilot flying in Europe. In 1944 Eugene had been declared Missing In Action along with his entire crew. His mother definitely didn't want her last son to be flying.

Joe Bossi's crewmen were the only Naval men undergoing training, since Taylor's men were already veterans. Bossi's gunner was the very young Burt Edward Baluk, a native of New Jersey. He had only just graduated from High School in 1944 when he joined up in the Naval Reserve. He went to Naval Air Technical Training Center Memphis, Tennessee, and then to Naval Air Gunner School at NAS Miami. He had been in the Navy just over a year when he was transferred to Fort Lauderdale with the rest. He had not been seasoned by any kind of combat, and at 18 years of age he still had the appearance of a freckled high school teenager. He was gregarious, levelheaded and loved music. He, too, was already awaiting his discharge. Civilian life did not pose worries for him. He knew he wanted to be an architect. A year in the Navy did him some good, but it didn't change his long-term goal to pursue a career he wanted since high school.

Herman Thelander, Bossi's gunner, had only been in since September 1, 1944. He was a young boyish sprout from Minnesota's Clap Township. For his entire enlistment, however, he had been stationed in Florida. He trained at the NAS Jacksonville, Florida, from September to November 1944, and he had been at Fort Lauderdale NAS from April to June 1945, before being transferred to NAS Miami, from June to August 1945, and then back up to Fort Lauderdale when Miami closed. He was a 4.0 student and a sharp, keen gunner.

The muster had been set, improvised, and short one man, but it was a good compliment. All were qualified, with some of the men like Thelander, Paonessa, Gruebel, "Harmon" and Parpart old hands on the training runs. Some like Gallivan and Thompson were indeed men of the "Gallant Hours" in the South Pacific.

These hours now yielded to the "quiet hours" as the crews silently endured the dusk flight west. Each man understood the controversy. Each knew to keep quiet and simply wait and they would soon be home.

Chapter 8

Fishbowl of Ink

GRAY LIGHT FADED; THE SEA TURNED CHARCOAL. DUSK WAS an ominous dye spreading throughout it. Soon, very soon, it would be black. Only the whitecaps would be faintly visible, flitting like disembodied spirits over that "ink bowl" of blackness, like a galloping army of the condemned coming forth to scour the landscape. The surging whitecaps said everything to the pilots and crew. Indeed, they said too much. They shouted the wind was coming from the southwest. But they whispered death. Ditching would be fatal in their grasping clutches.

The western horizon hailed the coming shadow. Darkness drained out all the color of life; only the gray mass of approaching scud— that gray inverted sea of clouds— carried any vigor of warmth from the distant setting sun. Only a few slivers of light streaked up beyond the westward scud; pink sunbeams spiking into the sky like the hand of a drowning man grasping at air before the final plunge.

The dome of clear sky within which Flight 19 had been flying was now constricting quickly within the vise of darkness. The scud was rolling overhead like a shroud being pulled by an invisible hand. Soon it would be black with darkness— a unique kind of darkness that only haunts a moonless ocean. Darkness at such a moment was a choking, claustrophobic hood. The dim glow of the control panel indicators would be the only light.

It is not hard to imagine the collective mood of the pilots and crew; it was heavy and not conducive to much banter. Anger over a routine mission having gone haywire and fear one plane might have to ditch were powerful motivators to induce an acrimonious silence. What Powers had done in taking command had to be done, but it was nevertheless distasteful. Taylor was no doubt feeling dreadful, and his rocking compasses did not make him feel any better.

The gunners weren't needed at their posts anymore. They could easily slump down into the narrow belly of the plane and sit on a fold-down bench. Gallivan could watch "Yo-Yo" Gruebel, the priest wannabe, remain enthusiastic. "Yo-Yo" would always have a word of comfort. Gallivan would be the sturdy, hardened Marine, wondering why he didn't remain an armorer. He wasn't scared. He was probably just peeved and anxious to get back. His last mission! What a foul-up!

One can well imagine what poor "Bob Harmon" was thinking. He had sweated out a rookie pilot months ago. Now he was in the midst of another harrowing encounter, this with a trained and seasoned pilot! Aside from being anxious to get this snafu'd mission over with, "Bob" was probably unaffected by the growing darkness. Ironically, aside from Taylor, he was the only crewmember who was truly qualified to be out there. Undoubtedly he contacted Taylor over the intercom and informed him of his training and duty off the carrier *Enterprise* during the war. Harmon would tell him how dark it will get, and how it will look like "a fishbowl of ink." Taylor might even relate it to the others to reinforce the potential dangers of the coming darkness. Such a simile might even strike the often-jovial Stivers as amusing.

Taylor, on the other hand, didn't need the advice. He could well remember flying over the Keys on nighttime patrol 3 years ago. He wasn't personally worried about ditching. But he now had 4 other pilots he felt responsible for. If he was alone, it would be easy for him. He had been in tighter jams than flying at dusk with a southwest wind. But what if one of the other planes gave out?

C.C.'s voice crackled calmly over the other pilots' receivers. It held a firm, instructional tone. It was 5:19 p.m. "Planes fly close formation. When first man gets down to ten gallons of gas, we will all land in the water together. Does everyone understand that?"

Each one picked up his mike and said "Roger."

Did he have to bring that up? Stivers was never too intimidated, not on the outside anyway. He was too much of an optimist. But he was not fighting an enemy. This was a question of skill. Taylor's instruction reminded him of the sea. Leaning forward and looking at those hungry whitecaps he could see that army of spirits. Powerless predators until one came down to their realm. What looked like small whitecaps now could be a mass of foam spitting up from mountainous, pitching waves. Now the waves seemed small, the whitecaps and frothing streamers like little confetti. But they concealed deep troughs and charging walls in that "fishbowl of ink."

They closed their canopies. It wasn't just to shut out the growing cold. It was also just in case they would indeed have to ditch in an emergency. No pilot wanted to ditch with his canopy open. It could slam shut and jam, and then where would one be?

Each pilot reaffirmed with his eyes the place of every key switch and lever on the dash panel. Altimeter and flight level would remain the ones watched most, as would be the compass and, of course, the fuel gage as it steadily slumped further and further toward empty.

The grayness of dusk was fading faster and faster. Soon they would be flying blind. Even in a squadron there would be loneliness. Each pilot was no longer visible to the other. The pacing Avengers were only dark silhouettes against an indigo sky at the verge of night. The green and red navigating lights were becoming more noticeable. These would soon be the only clue that another plane was still in formation.

All the pilots had been trained in flying blind. The routine played on in their minds. They could remember the drills. They were blindfolded while they sat in an Avenger on the tarmac. The instructor stood on the wing with his notepad and asked them to touch this lever, locate that button, turn this knob, flick a specific switch. They had all passed, of course. That was required before even getting to this level of advanced over-water training. But this was nevertheless a baptism of fire.

Taylor's voice came over their receivers again, calm and confident. He told them to stick close, watch the navigation lights and keep their eyes glued to their equipment. "What time is it?" he then asked again. Bossi perfunctorily responded.

Chapter 9

Destination Indigo

As darkness fell, Operation Radio's small square room seemed to constrict. It was the only floor in the tower from which lights shown forth, glowing into the dusk to reflect the anxiousness and attentiveness of the men inside. They hung closely on the radio speakers, but heard little if anything coming from the flight. Static and faint sporadic blurbs were frequently drowned out by Cuban music. Spanish disc jockeys announced the call letters with their lively Latin-Caribbean enthusiasm. Calypso surged over the waves. Sentences were suddenly halted by a whine of the airwaves or staccato bursts of static, half words and cut off dialogue providing a comical interlude.

It was essential to get clearer reception. The frequency of 3000 kilocycles was the standard "CRASH" frequency which all of the Gulf Sea Frontier and Air-Sea Rescue facilities monitored. It was standard procedure and indeed regulation for any pilot to switch to that frequency when in trouble. The pilot could then be heard up and down the entire Florida coast, and could be in two-way contact with any number of rescue bases. The weak training frequency of 4805 kilocycles had a very limited range; only about 125 miles.

"Nan How Able Three to Fox Tare Two-eight, Nan How

Able Three to Fox Tare twenty-eight." Operations clearly heard the Port Everglades' operator. "If you can change to Yellow Band [3000 kilocycles]," he continued, "please do so and give us a call."

Operations— Don Poole, Robert Cox, everybody— waited.

Dusk brought a strange, cool repose at Port Everglades. It was silent overhead. Fort Lauderdale's flights had landed. The surf's methodical, calm roar and boom was undisturbed by growling motors. It brought that sleepy relaxation that comes with a steady rhythm upon a rocking cradle. The radio room door was open and casting its light into the quiet evening. Thompson leaned in the frame, watching the scud approaching from the west. The operator, turning in his chair around from his apparatus, somewhat cynically recited a new fuel estimate for the flight.

Yet another estimate? There had been one that said the flight would be out of fuel at 6:30 p.m. Another said fuel for 5 and one half hours. That would mean 7:30 p.m. Now what did this one say?— 7 o'clock! Thompson couldn't help but shake his head before he went back in and did some calculations. One thing was certain. The average speed of an Avenger was 140 mph. Just from rechecking the dialogue, which had been quickly penciled by the operator, it was plain to see the flight hadn't headed northeast for more than 22 minutes, and certainly didn't head east for any appreciable amount of time. If they had at most been 150 miles out to sea, as he had first figured, they would easily make it to shore. It was now 5:22 p.m. The flight should be back in about an hour. Boca Raton, with its 150 mile radius, should be picking up their emergency IFF soon.

. . .But where the hell is that radio fix? It's been over 20 minutes since it was "soon" to materialize.

Operations Radio at Fort Lauderdale was also getting anxious about that. No reply from Taylor to Port Everglades' request to switch channels was worrisome. Reception should be getting better as they continued west. Shouldn't it?

That damn Cuban music! Two minutes of broken Cuban calypso.

Then a faint: "Nan How Able Three, this is Fox Tare twenty–eight. . ."

All those in Operations collectively leaned in closer to the speaker. Yet nothing else came over except Port Everglades' operator in response. "This is Nan How Able Three, shift to 3000 kilocycles."

"I receive you very weak," said Taylor. "How is weather over Lauderdale?"

"Fox Tare two–eight, this is Nan How Able Three, weather over Lauderdale clear. Over Key West CAVU. Over the Bahamas cloudy, rather low ceiling, poor visibility."

Nothing but static then some jazzy Cuban music butted in.

Silence continued.

"Is that a ship on the left?"

It was Taylor's voice. It was louder than before. This stirred some excitement at Operations. If it was a ship then they were perhaps over the sea lanes off Florida.

The silence of static. The most infuriating sound.

"What time is it?" Taylor asked yet again.

It hadn't been long since he had last asked. His voice was clearer. He *was* coming closer to shore. He must be.

"We've been heading west for about 15 minutes," replied Powers. (Taylor asked at 5:31 p.m. Therefore it is safe to assume Powers stated as much.)

"Nan How Able Three, Can you hear me?" Taylor then asked.

C.C. had the same idea as those on land. He should be getting through better if he was coming closer.

The Port Everglades operator responded quickly. "Hear you strength three, modulation good."

However, when the operator un–keyed his mike that lousy static and pugnacious Cuban music burst over. An angry, impatient snort. He tried again. "Nan How Able Three to Fox Tare twenty–eight, Can you shift to 3000 kilocycles? Over." Nothing again. "Fox Tare two–eight, please change to 3000 kilocycles. . . shift to 3000 kilocycles! Over." (5:34 p.m.)

No response.

It was plain enough at Fort Lauderdale that the weather forecast wasn't going to hold. The wind was increasing. The gray scud was slowly rolling over the western horizon and its menacing lip was right on the outskirts of the city.

The frustration in Port Everglades' operator's voice was evident to those in Operations Radio. Poor communication had been going on now for almost 2 hours. First Cox had trouble and now the reception still wasn't getting much better. Just where had this flight been and just where in the hell are they now?

Radio fix be damned! Poole was ready to send the Ready Plane. There had been no word from Baxter and there was no time to lose. He was giving orders to send it on a direct easterly course when a special message came in from Aerology. Palm Beach just transmitted that overcast and scud clouds were there and rain might be soon coming. That clinched it. Poole ordered the Ready Plane to stand down.

Cox still quietly clung by the door. He knew a great mistake had just been made and this caused his stomach to tighten up like a fist.

★   ★   ★

Thompson had waited by the radio, sure he would soon hear Taylor's voice coming clearly over 3000 kilocycles. Yet as the void now strangely continued he thought it time to contact NAS Banana River (modern-day Cape Canaveral) up the coast and ask them to try to raise the flight on its current frequency of 4805 kilocycles. In the pecking order of hierarchy, Banana River was a major hub in the Air-Sea Rescue web, right under Miami in the chain of alert. In any emergency Port Everglades would actually inform them first. They in turn would call Miami and have all things coordinated.

Thompson was having his message keyed into the teletype when Taylor's voice finally crackled over the receiver at 5:38 p.m. C.C. was still trying to feel his progress closer to the bases by how his reception was improving.

"Nan How Able Three, How do you read?"

The response wasn't good. "Very Weak. Change to 3000 kilocycles," replied the operator.

"Hello Nan How Able Three, this is Fox Tare twenty-eight. I can hear you very faintly. My transmission is getting weaker."

There was nothing for 10 minutes. It was 5:50 p.m. now. They had been heading west for at least 35-40 minutes. They should have made at least 75 miles progress toward the beach. Why wasn't the communication better?

Taylor's faint voice broke over again. "Hello Nan How Able Three. This is Fox Tare twenty-eight. Over."

Port Everglades' operator wasn't going to waste words. "Change to Yellow Band, Channel One, 3000 kilocycles and give us a call," he replied.

"My transmission is getting weaker."

What did that mean? Was he losing power? Were the other pilots telling him they weren't picking him up as strongly anymore even though right next to him?

Radio was not proving to be the best guide for the flight. That much was clear to Operations Radio. Hines called Miami's Air-Sea-Rescue Headquarters to request that they alert all airfields from Miami to Banana River and have them turn on their base lights. He then called Thompson at Port Everglades so he could broadcast this to Flight 19.

The time was now 5:51. p.m. The wind moaned past the radio room windows at Port Everglades. It was the only sound Thompson could hear after he hung up the phone with Hines. Its haunting tone westward carried a frustrating irony. These baby breezes would soon reach Flight 19 out there, somewhere, as mature gusts. Yet the clear messages being sent by the radio operator merely dissipated before reaching the lost squadron.

"Change to Yellow Band 3000 kilocycles and say words twice when answering," the frustrated operator repeated.

His words were almost stepped on by the teletype machine's mechanical processing. Hines request was being sent out— "Miami to all Air Sea Rescue Stations. All stations Banana River south light up to greatest possible extent."

5:54 p.m.

"Nan How Able Three to Fox Tare two-eight, Did you receive my last transmission? Change to Channel One 3000 kilocycles."

"Repeat once again." Taylor's voice was a muffled holler.

Possible bad weather approaching, hell! This was an Air-Sea-Rescue post. Thompson got on the phone to Dinner Key,

Miami, over the command line. "Request you send out Miami Dumbo to try and intercept FT-28 on 4805 kilocycles. We suggest that the Dumbo come north taking a departure from Fort Lauderdale and steer course 045° for 150 miles attempting to establish communication with FT-28."

5:55 p.m.

Trying to control his frustration, the operator again called over his microphone. "Change to Channel One, 3000 kilocycles."

Taylor emphatically blurted back: "I cannot change frequency. I must keep my planes intact." Calmly followed a minute or so later by: "Cannot change to 3000 kilocycles. Will remain on 4805 kilocycles" just to make sure he was being picked up.

It was now 6:02 p.m. Another pilot's strained voice blurted over the radio. Port Everglades heard: "We may have to ditch any minute!"

What was going on?

6:03 p.m.

The High Frequency Direction finding operator at Pensacola was surprised that of all people he actually now picked up a line. It reached out of the dark and struggled through the static. "Hello Powers, do you read me?" Pensacola was in the Gulf and the last they had been told was that the planes were lost off the east cost of Florida. He gently played with his tuning knob and listened closer.

"Hello Powers, this is Taylor. Do you read me? Over."

"Roger. I read you."

Pensacola now received surprising news. Houma, Louisiana's HF Direction Net, reported they had picked up FT-28 talking to the tower. Pensacola quickly clarified that it was not the tower but that the other pilot's name was Powers. They both listened carefully and jotted down as much as they could understand.

"Hello Powers. I have been trying to reach you." Taylor sounded a little irked.

"I thought you were calling base."

"Negative. What course are we on?"

"Holding course 270."★

"Affirmative. I am pretty sure we are over the Gulf of Mexico. We didn't go far enough east. How long have we been on this course?"

"About 50 minutes."

"I suggest we fly due east until we run out of gas. We have a better chance of being picked up closer to shore. If we were near land we should be able to see a light or something. Are you listening? We may just as well turn around and go east again."

Snippets and no more is all that Port Everglades could hear, but Pensacola and Houma picked it up almost complete, enough to make it possible to offer logical fill-ins (underlined) where the audio cut out. "Holding course 270" was clear enough for Port Everglades. But Taylor complaining about heading east sent shudders through Thompson. It was still somewhat of a mystery to them whether Powers had actually taken over the flight. Taylor's response, however, subtly qualified it; but only Houma and Pensacola had picked up the word "suggest," indicating Taylor was no longer in a position to give orders. Port Everglades could only pick out "Are you - - -" and then "We may as well turn around and go east again." Believing Taylor was ordering them east caused confusion and alarm. That would send the flight into certain, agonized death at sea.

Port Everglades therefore tried yet again to raise the flight. "Nan How Able Three to Fox Tare two-eight, do you read me? Ten, nine, eight, seven, six, five, four, three, two, one. Can you read me?"[10]

Nothing.

A few minutes later at 6:13 p.m. Thompson had the operator contact the crash boat, for all that might be worth. The operator slid the mike forward and almost kissed it. "Nan How Able Three to Nan How Able Four-One, Nan How Able 41: Switch your TCE to Blue Band and see if you can hear Fox Tare two-eight with a flight of five planes x-ray Give him a call on 4805 kilocycles and see if you can contact him x-ray.★" He waited

---

* Only Port Everglades heard Powers' reply, but thought it was Taylor.

10 at 6:07 p.m.

* Spoken to indicate a period.

for a response.

For just a brief moment, the radio whined and hummed. The Cuban music faded and Taylor's distant voice came over.

"Hello, Powers?" Two minutes later— 6:17 p.m. "Powers, what is your course?"

[Powers did not reply]

At this very moment there was triumph at Dinner Key. All stations had come in with radio bearings on Taylor's broadcasts. They quickly computed them. Due to the weakness of the transmissions, and considering how broken and far apart some of them were, it was not precise. But it was now triangulated. They could say that at 5:50 p.m. Taylor had been within a 100 miles radius of 79 degrees West and 29 degrees North— in other words within a radius of about 100 miles off New Smyrna Beach *in the Atlantic*. They had him!

This crucial fix, however, was already 30 minutes old at 6:21p.m. when Miami finally sent it over the teletype to Banana River. En route, Port Everglades intercepted it. Finally they had the fix! Thompson tore it from the teletype and had a quick "Roger" sent back.

Fortunately Thompson knew the teletype line northward from Port Everglades wasn't working and that Banana River could not have gotten the message. He immediately got on the phone and relayed Dinner Key's message, adding: "Please have all stations in this area attempt to contact FT-28 on 4805 kilocycles and advise him to maintain course of 270 degrees. This is serious now since planes have been lost since 4:21 p.m. and their gas should be out by 7:30 p.m. The teletype system is out north of here. Also, alert dumbos in your area."

Thompson then informed Dinner Key: "We called Banana River by telephone and had them alert all fields south of Jacksonville."

Miami was already ahead of the game. With a general fix now in, HQ ordered the Dinner Key dumbo to take off. It was airborne at 6:20 p.m. and darting out right toward the last radio fix position. Its instructions were to guard 4805 kilocycles and attempt to contact the flight.

But the airwaves had gone silent again. Port Everglades, Fort Lauderdale, Pensacola, Houma, none could hear anything now.

It was not until 6:37 p.m. that one faint line was heard by Pensacola, and yet it was the exact thing Taylor had last been heard to ask. "What course are we on now?"

Then a few minutes passed. For the first time Fort Lauderdale was able to pick up the call letters of another plane in the flight. It was FT-3— Bossi. His young voice was calling to FT-28. Presumably, since Powers was not responding to Taylor, Bossi was acting as the go-between. They continued to talk at few minute intervals from 6:44 p.m. onward, but Operations Radio at Fort Lauderdale couldn't understand any of it.

When it fell silent, Port Everglades quickly tried to get Bossi.

"Fox Tare Three, Fox Tare Three, This is Nan How Able Three. Come in please. We are reading you very weak. Come in please."

At 7:04 p.m. Bossi's young, strained voice came over briefly: "Fox Tare Three, Fox Tare Three, Fox Tare Three. . .to Fox Tare two eight, Fox Tare two eight. . ."

Then, finally, it fell silent again.

Operations Radio heard no answer. Port Everglades didn't even pick up the last line where Bossi called Taylor. Pensacola and Houma couldn't read a word either. No one would ever hear a word again.

Chapter 10

Scramble!

UNLIKE FORT LAUDERDALE'S OPERATIONS RADIO AND PORT Everglades' Air-Sea-Rescue radio room, Banana River's ASR room was free of frustration and tension. The reason for this was simple. So far, they hadn't heard much of anything. "They" is actually a grandiose pronoun in this context. "They" actually boiled down to Lt. Floyd Parmenter. He was the Air-Sea Rescue Duty Officer that night; he was the *entire* Air-Sea-Rescue department at this time. He sat alone in his small office adjacent to the base's Operations Duty Office. He did odds and ends and stood his boring watch.

Being short staffed didn't seem critical until this moment. Lt. Duane Walker, the base Duty Officer, came rushing in from Operations next door. Alarm and urgency was in his voice. He told him that the flight was actually lost approximately 150 northeast of the Bahama Banks with 30 to 60 minutes of fuel left.[11] The Dinner Key dumbo was already heading toward that position. The lost planes were on 4805 kilocycles.

11 They were thought to be 150 miles from base, over the Bahama Banks, not 150 miles northeast of the Bahama Banks.

Parmenter was flabbergasted. Messages had been coming into Parmenter via Walker since 5 p.m. But all Dinner Key had
first said was that some planes were in trouble *near the Keys.*
Then at 6:10 p.m. Parmenter was told that the planes were 5
TBMs out of Fort Lauderdale. Now he was being told that they
were actually lost *north of the Bahamas.* From what he first had
been told it sounded like Banana River only *might* get involved. But now this was being thrown right into his lap.

Walker quickly ducked out to get the Duty phone. Parmenter was on the Air-Sea-Rescue line to NAS Vero Beach,
south of Banana River, and had them ready their dumbo. At
6:17 p.m. he called Canova Beach radar station (also south of
Banana River) and alerted them there may be an incoming
flight. A few minutes later he alerted the 104-foot crash boat at
Banana River to standby. He then called the Air-Sea-Rescue
Task Group #3 at Fort Pierce and had them and their boat
stand by. At 6:30 p.m., he informed the Banana River tower
and asked them to please guard 4805 kilocycles. Manning the
Tower was Seaman Tramberg. He made it clear they could not
transmit on 4805 kilocycles, but they could listen.[12]

At 6:36 p.m. Walker leaned back in the doorway. "Port Everglades just called and said the coordinates of the flight are 79
degrees West, 29 degrees North."

"That's to the north!" he cried.

Parmenter was amazed he had not been told sooner. He
immediately called NAS Mayport, north of Banana River, and
told them to dispatch the Mayport dumbo. Within minutes he
instructed NAS Daytona Beach to do the same.

Dumbos, the nickname for the PBM Martin Mariner, were
huge "flying gas tanks" carrying a crew from 10 to 13 men.
These aircraft were designed as the long-range scouts of the
fleet and also doubled as the main rescue craft since they could
land and take off on water. They were, essentially, flying boats.

---

12 Until this time a major hub of Air-Sea-Rescue had not even been listening for the
flight. The tragedy of this was that the radio position fix showed Banana River was one
of the closest bases to the flight at that time.

While all these bases had a dumbo, Banana River was the main training base and had a fleet of them.

But at such short notice Parmenter was caught shorthanded. Lt. Walker rushed back into ASR and told him he had to request planes from the training squadrons. But there was a snag. "Was it necessary to have flight instructors on the planes?" he asked, adding, "Do you want me to call Commander Lawrence?" (the base commander).

Parmenter quickly answered yes to both questions.

At that point Lt. Commander Norman Brule called down to Lt. Johnson at the Duty Office on the squawk box and held back the crews of two big Mariners awaiting takeoff for a training exercise. They were Training 49 and Training 32. Brule told Johnson to find two qualified and fully certified pilots— quick! By a stroke of luck a topnotch instructor, Lt. Walt Jeffery, was in the office. He quickly volunteered for the mission and went to get his things. By 6:40 p.m. it was discovered that the instructor for Training 32— Lt. Gerald Bammerlin— was actually scheduled to be aboard.

Along with their student officer pilots, they gathered in the Duty Office for briefing. Lt. Johnson was tracing the area on the map with his finger as he spoke. Bammerlin was to takeoff and fly direct to 29 degrees North by 79 West to conduct an expanding square search. Jeffery was to fly north up the coast to 29 degrees and then east to 79 degrees. When there, he would contact Bammerlin and they would coordinate their square grid search.

"Both planes are to use radar and keep lookouts posted throughout the plane. Send back hourly reports to base on 4385 kilocycles. Guard 4805 kilocycles, the voice frequency of the TBMs. Contact the tower on 3065 kilocycles for updates of search information. That's it, gentlemen."

The group quickly dispersed, the student pilots heading to their planes while both Jeffrey and Bammerlin briefly examined the weather forecasts and then contacted Aerology for any updates. Because of the nature of the search and the prevailing scud, both were going to fly very low over the ocean. They left the Duty Office around 6:50 p.m. and headed to their planes.

Bammerlin's plane went "over the side" (rolled down the ramp into the water) first. His big gull-winged PBM was now chopping up spray in the narrow, protected Banana River harbor while Training 49's men were boarding their plane.

While they were getting to their places and stowing their gear, Maintenance Department machinist Clarence Urgin was still aboard inspecting key parts of the aircraft. Because the flight had been changed to a search mission, he went over the aircraft with a fine toothcomb making sure all was operational. Now at 7:04 p.m., just as Bammerlin took off, he had spent a little over 30 minutes checking Training 49 out.

Walt Jeffrey was up in the dimly lit cockpit with Harrie Cone, the student pilot. "Training number 49 calling radio NAS Banana River."

"Go ahead."

"Should I send a "V"?

"Go ahead."

Jeffrey was testing his radio strength. In response to his message, Radio Banana River responded: "Strength five, good."

"Roger," responded Jeffrey.

They were strapping themselves in when Urgin popped up between their seats. "All is A-1, no problems, skipper. All engines tested normally in pre flight, no fumes. This is the first flight of this bird after its 90-day inspection, so it's all tip-top."

"Great. We've got everything stowed, and we're under weight. We're ready."

"Have a good flight."

Urgin got off at 7:15 p.m. and immediately had his ground crew put 49 over the side into Banana River Harbor. The clumsy looking dumbo slid in and wallowed back and forth until steady. At 7:30 p.m. Jeffrey sent the message they were airborne.

Bammerlin and his crew had grown impatient circling over the patchy lights of the base for 30 minutes. Now they were finally relieved when 49's blip showed up on their radar. Bammerlin had good cause to be anxious. He had been told at 6:30 p.m. that the TBMs only had 20 minutes of fuel left. This meant they were already down.

When they finally paired, they headed north along the coast. Then, according to orders, Bammerlin broke off and darted out to sea while Jeffrey continued north to 29 degrees North Latitude, where he would then head out to sea and approach the position fix from a northern direction— just in case the flight continued to be blown north before it ditched.

The search had finally really begun.

Two PBM Martin Mariners in flight off Florida, 1945. The second PBM is the actual Training 32.

Chapter 11

The Drunken Frog Hunter
and other
SNAFUS
(The Search for Flight 19, December 5 – 10, 1945)

THE FIRST SNAFU OF THAT DARK NIGHT CAME WITH THE orders accompanying Bammerlin and Jeffery. By the time they took off Flight 19 could have been as much as 150 miles beyond the area of the 5:50 p.m. position fix. It was prudent for Miami to have ordered the Dinner Key dumbo out to that spot at 6:20 p.m., when they first heard of it, but by 7:30 p.m. there was little chance that the flight was still within that 100 mile radius. Nevertheless, five dumbos— from Vero Beach, Mayport, Banana River, and Dinner Key— were now converging on a fix that was terribly obsolete.

Commodore Howard Benson, head of ASR, had opened up the search by issuing an all points bulletin at 6:56 p.m. Alerting all ships to be on the lookout, however, is not the same as a diligent search. There was only one real search presently ongoing and that was limited to the obsolete fix.

To make matters worse, Dinner Key's dumbo had essentially disappeared after takeoff, with no base able to reach it. About an hour afterward they finally contacted Miami and informed ASR that their antenna had frozen over. Thus, ironically, the one flight that would have had a genuine probability of overhearing Flight 19 had been incapable of hearing anything or contacting them in turn.

The problems with the dumbos continued. At 8:30 p.m. Training 49 had not sent its hourly position report and nobody knew what had happened to it.

Yeomen rushed about the "bridge" at ASR relaying teletype messages as Benson, Commander Richard Baxter, and Lt. Commander William Murphy walked around the large map table. Meanwhile teletypes hummed and pecked in the background and the phones continued to ring. Over the radio a monotonous phrase kept repeating. "Fox Tare 28, Fox Tare 28, this is NSO, NSO Miami, how do you read me? Over." It seemed a useless gesture that Miami's NSO radio was still trying to raise the flight, but while ASR waited for some positive news it was the only active thing to do.

Only one report had so far come in, and Commander Baxter, Benson's chief assistant, and his staff were still busy trying to interpret it. It had been received at 7:50 p.m. from the escort carrier USS *Solomons*. At 7 p.m. her radar had picked up a flight at 29 degrees 35 minutes N and 81 degrees 28 minutes W consisting of "4 to 6 planes. No IFF. . .course 170 degrees; speed 120; estimated altitude 4000 feet."

The coordinates placed the flight right near Flagler Beach, Florida. But the direction indicated they were heading south, following the coastline. Baxter dismissed it. He assumed that by now had that been Flight 19 they would have seen and landed at NAS Daytona Beach. With all the bases lit up to maximum extent, Baxter's attitude to dismiss this report seemed a logical reaction.

But it is a fact that at that altitude Flight 19 would have been embedded in the heavy scud and unable to see the bases or even the coast.

By 8 p.m. the storm was already far out to sea. Lt. Gerald Bammerlin's dumbo was now over 29° N 79° W, where he reported that the ceiling was at most 1,200 feet. Winds were west/southwest between 25 to 30 knots. As a result the sea was very rough, covered with surging whitecaps. Within this narrow margin, his PBM pitched up and down.

Turbulence was so rough that it was a chore for the radar operator to closely monitor his screen. Numerous targets blipped onto the scope, each causing Bammerlin to break off his routine grid pattern. And over each he bravely made a low pass at 400 feet altitude. Each, however, would inevitably turn out to be yet another fishing boat bobbing in the frothy seas, either heading back for the coast or hauling in its nets. Bammerlin also buzzed several freighters, all proceeding on their courses. Columns of deep gray haze meant rainfall on all quarters. Often heavy drops pelted and streaked his windshield, for all intents and purposes causing the hearty searchers to fly blind.

At 8:45 p.m., Lt. Parmenter finally got some help. Lt. J.J. O'Neal reported for duty. At least Banana River's ASR headquarters now consisted of two men instead of one trying to figure out what was going on. Ironically, aside from intercepting messages, ASR HQ Miami, though fully staffed, was mostly dependent on what Parmenter was finding out.

The second great SNAFU occurred at 9:10 p.m. WAX Miami intercepted a dispatch from the freighter s.s. *Gaines Mill.* "At 7:50 p.m. observed burst of flames, apparently explosion, leaping flames 100 ft. high burning 10 minutes; position 28° 59 minutes North 80° 25 minutes West. At present passing through big pool oil at 8:19 p.m. Stopped, circled area using searchlights and looking for survivors; none found."

Baxter and Benson calculated Flight 19's flight track from its 5:50 p.m. position fix. With the winds that had prevailed, it was more than likely they would come into the coast somewhere north of Banana River. Could the explosion have been them blowing to pieces in a spectacular midair collision in the rough turbulence? ASR ordered Bammerlin to head in to the area.

Bammerlin was more than ready to go. Training 32 had been on a roller coaster ride of turbulence. The wings were flapping up and down stiffly like an arthritic seagull. Coffee shot out of cups; it was nearly impossible to drink or eat (not that anybody truly wanted to). Bammerlin gladly pulled back on the stick and turned to port.

Due to the strong 30 knot headwinds, it took him about an hour to get to the spot. The low ceiling made the area claustrophobic. Between sea and scud there was only about 1,000 feet of clear sky to fly in. At 800 feet altitude it wasn't

The first big SNAFU. The small circle is the 5:50 PM position fix. By 8 PM that night (the final fuel estimate) Flight 19 could have been anywhere in a radius of 200 miles (large circle) of that fix. None of the search pilots were ever told this.

hard to spot a 100-foot boat bobbing up and down in the surging sea near what looked like an oil slick. Bammerlin was sure it must be the Smyrna crash boat at the scene. He branched off from here and started his square search. Again, numerous fishing boats, a couple of yachts and several freighters were overflown, but so far nothing that looked like debris, rafts or bodies in the pitching seas.

A new SNAFU now began. The search had initially been misdirected to the old position fix report, but now passing freighters were reporting a show of flares. This forced ASR to start playing checkers with the PBMs. The first report came in almost 2 hours late at 10:24 p.m. It was from the s.s. *Thomas Payne* off Cape Canaveral. The report was actually timed at 8:35 p.m., but ASR nevertheless diverted the Dinner Key dumbo. The next message was from the s.s. *President Tyler*, reporting "apparent aerial white flare bearing 100T[rue] from position 29.30 N and 79.40 W sighted at 11:30 p.m." This was just being handed to Lt. Commander Murphy, Baxter's assistant, when another message came in. It was from the s.s. *Jeremiah O'Brien*: "At (GST) 11:30 p.m. sighted flare to inshore searching vicinity approximately position 28-50 N and 79-40 W." In reality, the Dinner Key dumbo, Training 32, the Daytona dumbo and the Vero Beach dumbo were chasing each others' tails. They had been dropping flares to illuminate the surface of the ocean. Each time a freighter reported these, another dumbo was diverted to search the area. The only exception to the rule was Training 60. It, too, had now vanished.

December 6 did not start off good. The Mayport dumbo was back at base to refuel and was going to remain down because of the weather. (The Mayport dumbo hadn't even been out for 6 hours yet.) There were precious few search eyes out to sea now, and ASR, still unaware that the dumbos were chasing their own flares, wanted more reports investigated. Bammerlin was still hugging the area of the 7:50 p.m. explosion (the Daytona dumbo was tracking down Bammerlin's flares). Both of the Coast Guard cutters *Vigilant* and *Pandora* were only now slowly steaming out into the rough seas.

A large changeover of personnel had occurred at midnight, but there wasn't much enlightening to tell them beyond 'five planes had to be down off the Florida east coast.' The map marking where flares had been reported looked essentially like the result of a game of tick-tack-toe.

To Bammerlin's surprise, a radar target they had been watching now turned out to be the long lost Training 60. It was 30 miles off Canaveral, tracking down flares as well, probably the Dinner Key dumbo's. With another PBM in the vicinity, Bammerlin informed base he was going to fly back out to the 5:50 p.m. radio position report. From there he would follow a line to the position of the explosion, assuming, of course, that this explosion had been Flight 19. By paralleling this route, he might see something. Besides, there was the report of yet another flare out there to seaward.

This was a rash act on Bammerlin's part. The seas were

pitching furiously and, except for the whitecaps, shrouded by the deadly dye of ink. Deep, black swells could conceal treacherous troughs, each one capable of swallowing his plane whole if he had to make a forced landing. Winds howled down, turbulence slapped them around, and the sea spit angrily up at them. In addition, the heavy rains, which were absent off the coast, began again and pelted the PBM heavily. In the midst of planes coming in and vessels being held back because of the weather, Bammerlin shot out into a very dark, tempestuous and lonely sea in what was truly the one act of heroism that night.

With no sightings of any flotsam at sea, Baxter began to wonder about that sighting of planes made by the *Solomons*. The hunch was a justified one, though a bit late now at 4:10 a.m.

With the hopes of quickly finding the pilots dashed, it was time to do a little backtracking. "Request any further information you may have on later position of planes," Baxter sent to *Solomons*. At this time the vessel was pitching out to sea further than any other rescue craft. She was chugging out to 77 degrees West Longitude, according to orders, just in case the flight went east.

While Baxter and ASR Miami awaited an answer, the original SNAFU reared its head. Reveille sounded at all the bases in Florida. Into crowded briefing rooms pilots and crewmen jammed to be briefed on the obsolete 5:50 p.m. position fix. The scene at Fort Lauderdale is typical. Lt. Commander Poole entered in full flight gear and stood before the chalkboard. He was brisk and to the point. He drew the area of the fix, scribbled the coordinates, and advised on wind directions and velocity. Fort Lauderdale was going to search the south end of it. It was a large area, but the strength of the signal didn't allow for a more precise fix. "Somewhere in here," he pecked the chalkboard with his chalk, "they must be. Look for debris, men in a life raft and bodies with Mae Wests."

Over 40 Avengers from NAS Fort Lauderdale droned over the coastline out into the Atlantic. It was a similar scene at every Naval Air Station along the Florida east coast. NAS Jacksonville sent out an armada of aircraft; some 70 in all droned out over the peaceful residences. Sleepy morning households were awakened by the rattle of fine china pieces, shaking windows and thunderous echoes. Even the Mayport dumbo raised the courage to sally forth with the rest of them. The Dinner Key dumbo was off at 7:45 a.m. Banana River was sending up more dumbos and PBY Catalinas. Vero Beach was launching again. A C-47 was off from Morrison AFB. The final count was 242 aircraft in the air.

Now, at 6 a.m., tired and spent and low on fuel, Bammerlin was heading back in. His radar operator just picked up dozens of radar targets heading out to sea. The vast search armada was going where they had just been all night. Bammerlin had stayed out to the fullest extent of his fuel. If there was one outstanding action during last night's hasty search, it was Bammerlin and his crew's heroism and attention to duty.

If any of the rescue crews— especially Bammerlin's— knew that the 5:50 p.m. radio fix was already over 2 hours old when the Avengers were estimated to have run out of fuel, they would have been enraged at how they had wasted time. In truth, Flight 19 could have been anywhere up to 200 miles beyond that spot when they finally went down.

This SNAFU only continued all day December 6. Every TBM and PBM flew at low altitudes over the choppy, gray sea, crisscrossing in elaborate lattice grid searches the area of the obsolete fix.

Coastal searches, however, produced the first exciting news. Banana River reported to ASR "Pilot of SNJ ready plane believed that he saw a body with a life jacket in position off Matanzas Inlet. Body was just below the surface of the water. King 10 is investigating assisted by blimp K118." Matanzas Inlet was south of St. Augustine and just a bit north of where the *Solomons* saw that enigmatic flight of planes at 7 p.m. last night. Could this be from one of them?

While various ASR bases waited for news, staff continued to go over much of last night's communiqués and in the process started to clean up some of the SNAFUs. Plotting the various PBMs' position reports, Banana River's ASR discovered that last night's flares were dropped by the dumbos themselves. Jacksonville NAS and Morrison AAFB had relayed messages that the explosion seen by the s.s. *Gaines Mill* coincided with Training 49's position. Now at 2:05 p.m. the news arrived that clinched its fate. It was from the carrier *Solomons*. "Our air search radar showed plane after takeoff from Banana River last night joining with another plane and then separation and proceeding on course and 045°. At exact time as s.s. *Gaines Mill* sighted flames and in exact same spot the above plane disappeared from the radar screen and never reappeared."

ASR Miami had little time to appreciate the tragedy of the news before a stream of negative news came in. At 2:13 p.m. it was confirmed that there was no trace found at Matanzas Inlet. Report was now evaluated as false, which meant nothing more than that they didn't find the stimulus of it. "We are securing this area."

Things were not looking good far out to sea either. Negative reports were coming back from the hundreds of aircraft and vessels. Then PT 614 reported heavy seas were forcing her to proceed to the beach at low speed. "Some damage sustained already." A few minutes later at 2:20 p.m. PT 613 sent a message: "We have also encountered heavy seas and one man has broken his ankle. We are forced to return to base. Request doctor in tender to be at pier 2 at 2:30."

By 4 o'clock that afternoon all reports were coming up negative. Most of the TBMs from the Naval Air Stations had already landed back to base, and those pilots who were from Fort Lauderdale were especially disappointed.

The next morning— December 7— was marked by remembrances for Pearl Harbor. This was the first December 7th of a post war peace. This very day 4 long years ago the war began in a flash of destruction and chaos as the Japanese navy and air force attacked Pearl Harbor and wounded the US fleet there. Had it not been for the war none of these men would have been here.

One of the most salient but silent SNAFUs was, from our vantage point today, subtly revealed on this day. *Solomons'* 7 p.m. December 5 radar report was still bothering Richard

The location of flares reported throughout the 6-day search. The X marks the explosion seen by s.s. Gaines Mill.

Baxter. Last night he had asked her to report what type of radar she had. She now included in her message (at 11:20 a.m.) that she has SC red, which has about a 100 mile range. With this, plus her steaming time and ETA, it is easy for us to calculate her position. Most obviously this tells us the squadron of planes she sighted near Flagler Beach was right at the edge of her radar range. But more significant than this, *Solomons'* radar report casts real doubt on the idea Flight 19 ditched at sea. Not because that 7 p.m. flight could have been them crossing the coast. There's something else. Double-checking her radar range with her position shows that her radar was covering the majority of the 5:50 p.m. position fix radius and virtually any place Flight 19 could have advanced by 7 p.m. If Flight 19 was still at sea at 7 p.m. *Solomons* would have detected it on radar. The flight could not have turned around and headed east long enough to have escaped *Solomons'* scope by 7 o'clock.

It is hard to imagine I am the first one to notice this. I find it hard to believe that Baxter, Benson and possibly Murphy did not notice while plotting the radius of her radar range (if so they did) that she was covering the entire westward area of the 5:50 p.m. December 5 radio position report. It is impossible to try and figure out what exactly went through Baxter's and his staff member's minds this day. But there had to be some suspicions that the whole search might have been misdirected to begin with. Why ask the *Solomons* this late in the game what her exact radar apparatus was if they were not pondering that radar report near Flagler Beach? And in discovering her radar range they should have suspected Flight 19 was no longer at sea by 7 p.m.

They could have easily calculated the position of the *Solomons* on the big map at ASR with their plotting compasses. We know that at 6:45 a.m. December 6 she would be at 29 degrees 45 minutes North and 77 degrees West. She signaled she was making 12 knots. In 12 hours time (from 7 p.m. December 5) she could only have sailed 150 miles to eastward. Therefore at the 7 p.m. radar sighting she must have been within a certain track-line between St. Augustine and Daytona Beach, about 80 to 100 miles offshore.

It must have been, or should have been, a source of tremendous agitation to realize that the *Solomons* was only about 30 miles northwest of the 5:50 p.m. December 5 position fix epicenter at 7 p.m. ASR should have been kicking themselves now. Why had they concentrated their search in this area? Something was terribly wrong, and this oversight becomes one of the deadliest SNAFUs of the whole search operation.

Perhaps the Flagler Beach report was easy to dismiss on the face of it because the *Solomons* should have, logically, detected the flight at sea for quite sometime instead of first off Flagler Beach. The only answer is that *Solomons* was not monitoring her radar the entire time. But by 7 p.m. we know she was, and by 7 p.m. Flight 19 was still in the air. And despite the vague impression of Fort Lauderdale Operations' staff, Flight 19 did not turn around and go east just because Taylor "suggested" it. What was that flight which *Solomons* picked up on its radar along the coast if it was not Flight 19? No other flight of 4 to 6 aircraft was still up in the air.

Among the SNAFUs of that night is the radio position fix itself. It was imprecise to begin with, but why didn't anybody at ASR ask the question: 'Why was there no further radio bearings taken on the flight's progress after 5:50 p.m.?' The flight was still being picked up an hour and 15 minutes later around 7:05 p.m. It would have been a great help for ASR (and indeed for us now) if there had been

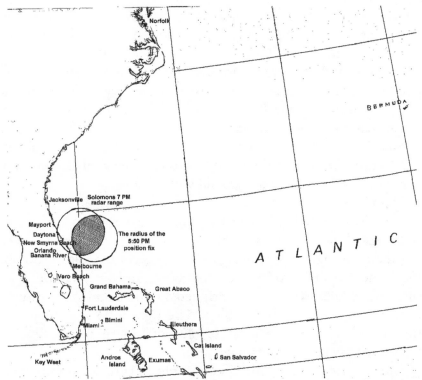

The extent to which the Solomons' radar coverage eclipsed the 5:50 PM position fix is marked in lines. By 6:15 PM Flight 19 would definitely have been within the shaded area.

follow-up position fixes on the flight. By 7 p.m. this would have shown either a westward continuance in their course or northward, eastward, wherever.

It is somewhat more reassuring to think that such questions were going through the minds of those present at ASR. The fact that Baxter and possibly also Benson requested to know the exact type of radar *Solomons* carried meant they were still trying to evaluate her radar report for 7 p.m. December 5. If that was not their motive, it is hard to explain why so much credence was given to what was soon to become one of the biggest SNAFUs of the search.

And this SNAFU surrounds the Florida swamps. On December 8 Eastern Airlines Flight 56 was at 3000 feet en route to Miami. Captain J.D. Morrison was at the stick. Next to him was co-pilot M. Gunther. It was 1:58 a.m. They were just near Melbourne, Florida, west of the Banana River area and south of Orlando. The fog hung here and there over the marshy terrain below. The lights of any settlement had long faded away. The dark waterways of the swampland flickered with the reflections of light off the gentle rolling currents. Little islands were everywhere, green vegetation, thin, spindly trees— everything that marks the swamps and bayous of central Florida.

Suddenly a small red rocket streaked up at them. It rose about 300 feet off the ground from a small island. It burst into a thousand sparklets and fell back to the ground with streamers.

"Jesus!"

Morrison pulled the wheel to port and began to circle the small island. Suddenly a searchlight was cast up and was following them. A man was running by the edge of the island, carrying some kind of strong search flashlight and was waving it at them frantically. Morrison continued to circle and lose altitude until he was about 300 feet high. The light followed the plane the entire time. As the plane leveled off, the person with the light ran to get a better vantage point.

Morrison turned on his landing light. There were hundreds of small islets in the area. The one the man was on was about an acre in size. The man started to wave the light again, this time directing its beam to the water to show where the plane could land.

Gunther took the stick as Morrison got on the radio.

The line from the Miami air controller rang at ASR at exactly 2:06 a.m. Lt. Covey picked it up. The Civil Aeronautic Authority was on the other end. "Eastern Airlines Trip 56 is about 10 miles west, southwest of Melbourne right now. At 1:58 a.m. a red flare with streamers was shot at them. They also observed a flashing light on land and are now circling the area." Soon Air Transport Command got and relayed an update. Morrison reported that he just saw a large fire halfway between Melbourne and Orlando, about 20 miles from the flare.

This inspired ASR to believe that these two locations could represent two separate Avenger crews. A Navy aircraft had quickly diverted and was already circling the area. The pilot quickly got on the radio to Banana River. "I just saw a red flare 15 miles inland bearing 245 degrees from Melbourne

NAS."

Melbourne NAS was unfortunately only on caretaker status. Nobody was at the base to go out and search. But now there was no denying it; it seemed like there were indeed 3 distinct locations where men were firing off flares or building a bon fire. This sounded like Flight 19's final resting place had been found.

Ground units from Banana River and Vero Beach rushed to their jeep weasels and amphibious craft. By 3:30 a.m. they were pouring out of the base gates (except for Banana River's weasel; its crew was left behind to beat and curse it because it wouldn't start). By 5 a.m. the rescue units took up their positions to enter the marshy swampland. Trucks were dropping scores of troops on the perimeter. The ambulance was waiting at the makeshift field HQ. The order was given. They moved in.

Fog was clinging to the ground. An incredible maze of waterways was veiled before them. Hampered by the heavy fog, the land units were still groping through only a fraction of the area two hours later. The weasel drivers stopped their motors periodically and the men listened. They could hear other engines, the amphibians and men sloshing through the marshes, but no cries for help.

Everything looked alike— every island, every marsh, and every strip of still backwater. All of it ghosted into view, reaching out from the heavy fog without warning: a ragged old dead branch, marsh reeds, a mass of brambles or a fallen tree. Searchlights only confused things in the heavy fog, making it look as though they were moving through a veil of choking, musty cobwebs.

Overhead somewhere was the Navy plane. Its orders were to circle the probable area as a marker. It didn't help ground units, though. Neither its engine could be heard nor could it be seen through the fog. By early dawn, no little island had been positively identified and no wreck had been uncovered. Nor could any of the units find any remains of a large bonfire. As the fog began to lift, boat motors could be heard gunning at greater speeds. Soon the still backwater channels were ebbing onto the banks in the wake of speeding motorboats. However, nothing could really be found. Ground units, exasperated, were calling Banana River. Exasperated, Banana River called Miami.

"We request you get from Captain Morrison an accurate position, either as a bearing and distance from a known object or in coordinates of latitude and longitude."

This inspired a short angry exchange between them and Miami. Miami sent back: "Are you cognizant of information in my first message?"

"We are cognizant of all messages received thus far on this situation. However, the position thus far does not seem to be definitely established by Captain Morrison."

"Position 210 degrees, 15 miles," Miami snapped back.

Over the phone Captain Morrison once again laid out his story in plain language to Commander Baxter. He volunteered to personally join the hunt. The problem facing Baxter was how to get him out there in the best way to do any good.

A couple of hours later, the ground units heard an unusual and new sound. It was the chopping of wind. It grew louder and louder and was descending over their heads. Pretty soon the brambly trees and seedy reeds began to swirl and wildly wave about. Over the treetops, a strange new thing came into view, beating up a fierce wind and rippling the still waters around the small islands.

The fog had lifted about 50 feet off the ground, and there Banana River's helicopter slowly lowered from the vapors like a divine gift to the searchers. The ground searchers saw a passenger inside pointing about this way and that. The Navy pilot and he were constantly conferring.

From the point of view of Morrison and pilot Winchell the men looked like ants scurrying about trying to escape the chopper's wash. Winchell pulled up the chopper and snubbed the men with a quick turn and lift of its tail. They were off and making good speed to another location.

It was now a veritable manhunt in the swamps around Melbourne. It swarmed with soldiers, jeeps, amphibians, weasels, and trucks. But unlike manhunts for escaped criminals, they were looking for men that presumably wanted to be found. But there was nothing. There were no more flares, no shouts of "Ahoy," and no holler by one of the searchers that they found anything.

Within this massive dragnet one person had escaped particular attention perhaps because he was in no trouble. Indeed, truth be known, he looked quite to be in an opposite state, very relaxed and content. He was a stocky man, a day or so of beard growth on his rather speckled, briny face. A begrimed t-shirt fit tightly over his taught belly. Over this was wrinkled an unbuttoned plaid shirt. His flat-bottom outboard was coming into one of the boat landings under the power of his lethargic polling. He sang, gurgled, and gargled a ragtime tune.

It turned out his name was L.C. Smith. Judging by the old sloshing tin pale of frogs that looked equally soused yet praying for intervention, he was what he claimed to be— a frog hunter. Once a hold of him, Winchell jogged his memory which, regrettably, often swam in 90 proof. Through resuscitating inhales and spinning eyes, Smith was able to describe something weird that had happened to him last night. Smith's thorax cleared a stale cloud of bourbon out of the way with a retching sound. He then described how a plane circled overhead. It came low and turned on its landing light. He thought the plane merely wanted to look at the surroundings. Yes, he had built a fire there. He was keeping himself warm while dead-eyeing his game of frogs. He couldn't understand why the plane was circling. He worked at the Conch Manufacturing Company, should they need more information from him. "What's all going on out there anyway?"

Well, Smith at least gave enough information for Winchell to give Banana River an estimate of where he must have been last night. Banana River, based on Smith's memory, evaluated Smith's position as having been 20 miles away from where the flare was reported by Morrison. Morrison, now waiting at Banana River where he had been taken, agreed. Banana River informed ASR that they will check further.

Miami also agreed. They couldn't believe it was a drunken frog hunter mad at an airplane for scaring his frogs away.

Morrison went back to Miami, where he and Baxter personally spoke. Baxter was soon talking to Banana River. "I have just talked with him [Morrison] and he is certain definitively of a few points which may have bearing. There definitely was no fire in the area that he circled after seeing the red flare. It was a marshy area. He did observe a large fire 30 miles or so further on while at 1,000 feet. He made three circles counterclockwise and the person was endeavoring to pass a message of some kind as to a possible landing on the water in that area." Baxter made sure more than Banana River received that message. He had Port Everglades pass it on to Fort Lauderdale.

Banana River checked further but was losing interest in the swamp search. A farmer had had a fire going all night near where he was tending his herd of cows, and Banana River thought this was the bonfire Morrison reported. They could find no trace of a soul out in those backwaters except, of course, for the redoubtable Mr. Smith. To them it seemed that Morrison really hadn't been sure of the exact place. Today's helicopter search had proven that to them. Winchell and Morrison had set down at several points but could find nothing.

There was little point to continuing the search, as far as Banana River was concerned. Their units pulled out, and by that evening the isolated swamp was peaceful yet again, except for the nightly orchestra of frogs. . .and perhaps again Mr. Smith's grateful liquid singing.

By the 9th of December, it was pretty much curtains for any hope the men of Flight 19 were in the swamps of Melbourne. In fact, by the 9th it seemed the curtain was slowly being wrung down on the entire search. It was unthinkable, but yes indeed 6 aircraft had vanished without trace. One could guess about the PBM, though why it blew up was a mystery. But for the 5 Avengers, what had happened?

The days after the "drunken frog hunter debacle," as it would later be called, were the same as those before. Freighters reported flares, and units went racing after them. The s.s. *Harold Jordon* spotted a large oil slick out to sea. A debacle at sea similar to the one in the swamp was caused by a report of two life rafts secured to each other with 2 men in them. Freighters and search vessels steamed for hours to get to the scene, dumbos raced to fly over and find it. The official conclusion was that it was two crates next to each other. A white flare was seen here or there. The s.s. *Irwin Russell* caused quite a stir when it was reported to have picked up survivors. This was later reported to be false. It had only sighted an old target. Reports of a parachute canopy at sea started another daylong routine of signals being sent back and forth. One dumbo even had mechanical trouble and had to "dunk" between two destroyers. The search craft themselves were falling apart from the daily use with no real maintenance work done in the hurried turnaround.

At dusk on the 10th of December it was over. The commanders, in consultation with Admiral Davison, the Advanced Air Training commander based at NAS Jacksonville, concluded there was no further reason to continue the search. A message was sent.

> Search for five missing planes of December 5, 1945, will be terminated on completion of return of all planes on this mission today. No further special search is contemplated. All planes and vessels in the area keep a sharp lookout and report any pertinent information.

Something must explain all of this. The largest search in history had failed to locate a trace of a flight which had merely been flying a routine training course. As the search ended, every man at every base in Florida knew there would be an intensive Board of Inquiry to try and sift out what had happened.

What nobody would know is that the Board of Inquiry would become the greatest SNAFU of everything associated with Flight 19. A silent SNAFU granted— for the records would remain behind locked doors for decades— but a SNAFU that would guide the Navy's actions and control its opinions for decades. Yet it would be a mesh of contradictions with only one goal in mind— who to blame. There was no investigation whatsoever. The most obvious went beyond the members of the Board. Loopholes big enough to drive a bus through were not closed. As the nation was stunned by the press announcements that the search was over with negative results, that 27 men and 6 aircraft had just vanished, the Navy red tape began.

Chapter 12

The Blame of Inquiry

ON DECEMBER 7TH, 1945, WHILE THE SEARCH WAS IN ITS SEC-
ond day, Admiral Ralph Davison, Chief of Naval Air Advanced Training located at Jacksonville, Florida, ordered the
Board of Investigation to be convened. Captain Albert K.
Morehouse, the commanding officer of Naval Air Station Miami, was to be its head. Comprising the Board with him would
be Commander Joseph T. Yavorsky and Commander Howard
S. Roberts; and Lt. Commander Richard S. Roberts would be
the Recorder. Davison ordered the Board to convene at NAS
Miami "at the earliest opportunity." The reason was clearly
laid out: "for the purpose of inquiring into and reporting upon
the circumstances attending the disappearance of five TBM
type aircraft. . .and one PBM aircraft . . .which occurred approximately 1900, 5 December 1945, resulting in the missing
status of Naval personnel."

After reading the Board of Investigation's hundreds of pages,
one must say, to put it politely, that a stricter interpretation of
these orders could not be conceived. It is difficult to write this
chapter and organize it, difficult perhaps because the Board
overlooked so much. To highlight the problem at hand, it is
necessary to examine the testimony of key officers in light of
the actual events of that December 5. However, I do not wish

the critiques and criticisms that follow to be confused with implying these questions and criticisms were going through the minds of the Board members and were just not expressed or that there was a "cover-up," as is always popular to claim today. On the contrary, after studying the voluminous records and testimony, I am sure the Board did not have a clue.

I must also make it clear that their oversights have not been uncovered by the "perfect vision" of hindsight. The inconsistencies that I noticed should have been readily apparent to each member. If they did not think of it during the testimony, they could do so afterward while going over the testimony for that day. It was within the Board's power to recall witnesses, which they in fact did on a few occasions, and put more questions to them and request more clarification in light of another officer's testimony. Most of these recalls, however, show they noticed nothing suspicious in the conflicting testimony, thus overlooking much of which was critical to a sound and objective conclusion.

The Board's myopic interpretation of their orders to merely set in place the circumstances cannot be used to justify the multiplicity of their oversights considering that they went way beyond their legal powers and decided to engage in an unauthorized mock trial in order to find blame. This seems to have become their sole purpose as a direct result of their inability to challenge obviously suspicious statements from some of the first officers called, especially those that cast C.C. Taylor in a negligent light.

The decision to maintain the Board at NAS Miami (although it had ceased flight operations it was still functioning as a Naval base) was caused by the usual Navy regulation to make sure there would be no conflict of interest in the officers presiding. Yet the officers who presided over it were stationed at either base, with Howard S. Roberts being the executive officer of NAS Fort Lauderdale, and the others stationed at NAS Miami. It is naïve to deny that they had a conflict of interest. All these men were disposed to vindicate their staffs.

If anybody should be in a position to instantly understand the criteria and detect any deficits, these were the men. But their ennui is self-evident in the 56 Facts they cite at the end of their deliberations, and the 56 Opinions based upon those dry facts. This was an opportunity for them to analyze all the testimony and to draw upon much of the discrepancies in the testimony of certain officers. This was a confidential report. There is no reason for them to have anticipated the birth of the Freedom of Information Act in the 1970s which would allow the results of their investigation and their synthesis of the testimony to be released. They could easily air their findings for Admiral Davison and his superior, Admiral F.D. Wagner, about clear incompetence without the worry they were giving the Navy a public black eye.

Justification for such a strong statement is found in the testimony of the very first witness, Commander Thomas H. Jenkins. He was Lt. Commander Poole's boss, and the officer in charge of the entire training program at the air station. Part of Jenkins' answer to Question 3 of the Board "State what you know concerning the loss of the five TBM Avengers. . ." should have raised eyebrows immediately with Morehouse, Yavorsky and Roberts. "The five aircraft involved comprised Flight No. 19 and took off from the main station field of U.S. Naval Air Station Fort Lauderdale, Florida at 2:10 p.m. on 5 December 1945. Flight No. 19 was due to return at 5:23 p.m., 5 Dec 1945."

It is with just those few words above that he introduced what is one of the most crucial foul-ups of the saga, and one which has remained completely overlooked until this moment. It alone would have been enough to court martial some of the officers. For in reality Problem Navigation No. 1 was a *two hour and 15 minute* flight not the *three hour and 15 minute* flight Jenkins' time slot gives it. This inaccurate ETA is not a typo or a misspoken statement. It is all over the communication logs sent between all the bases, and will be testified to by others.

The Board really should have suspected something since they were all pilots. If it was not immediately apparent to them, it should have been after hearing Jenkins recite the amount of distance of which each leg was comprised.

If this was not enough, the testimony of Lt. Willard Stoll days later should have been. When asked what his duties were on December 5, he replied matter-of-factly: "I was acting as instructor for a navigation flight for problem number one from NAS, Fort Lauderdale. We had Flight number 18 and we took off at 1:45 p.m. due to return at 4 p.m., 5 December 1945. We flew at 1000 feet so as to use surface winds." The true Problem Navigation No. 1's actual time slot— 2 hours and 15 minutes. Captain William O. Burch, the base commander, even declared this time period to newspapers, saying the flight time was "only a little over 2 hours." Flight 19 should have been 23 minutes behind Flight 18 at 4:23 p.m. The Board didn't even notice. This hour mistake would prove fatal to the members of Flight 19. ASR was not prepared to declare the flight overdue until 5:30 p.m., over an hour past the real ETA.

This is not a small point. The ETA foul up is key to the launching of the entire search operation. In fact, the malaise over launching the search is directly the result of this terrific gaff.

The consequences of the ETA gaff do not apply merely to the later drama of Flight 19. Lt. Commander Charles M. Kenyon's testimony mutely proved how the ETA error was responsible for Fort Lauderdale's ennui at responding to the flight in the beginning of its trouble. He was the second officer called to testify. He was the "Hands-on" Operations Officer. That made him the staple in the food chain of responsibility. When the Board asked him what he did after being informed the flight was reported lost he replied in part: "The first action I took was to send the duty officer [Hines] to Operations Radio. When I was first advised that these planes were lost, about 4:10 p.m., I assumed that it would be almost impossible for them to get into any serious difficulty, and did not go to Operations Radio myself at that time." Just because of those few words, the Board wanted him to "explain your basis for this assumption."

. . .I checked to find out what time the flight was due back and it was 5:23 PM. This transmission of planes being lost was picked up at 4:05 PM according to our radio log, so there was another good hour before they were actually due back which indicated that on problem one navigational hop they couldn't be too far off since a 20 minute bombing period is conducted at Hen and Chicken Islands. I figured they were temporarily confused and, with all the instructions we had given them and with another plane in the air that had happened to pick up the transmission, they would come back right on time.

Not only is it significant that Kenyon mentions the erred ETA, his calculations implicitly reveal he knows Problem Navigation #1 is roughly a 2 hour and 15 minute flight. With an ETA of 5:23 p.m. Kenyon must have thought the flight took off at 3:10 p.m. (Kenyon never mentions any time of takeoff). With this in mind we can understand why Kenyon thought they were still close. It took about 15 to 20 minutes to take off and get to Hen and Chicken Shoals. Then 20 minutes spent there to bomb a target. That's about 40 minutes of the 2 hour and 15 minute flight right there. Thus when Operations was told that a flight was lost on Prob. Nav. One around 4:05 p.m. with an ETA of 5:23 p.m., it logically could only have been 10 or 15 minutes east of the Biminis, which themselves are only 55 miles off the coast from Miami.

Had Kenyon or any other officer realized that Flight 19 was at 4:05 p.m. just 20 minutes away from their *real* ETA of 4:23 p.m., they would have realized the flight could have gotten lost at the extreme edge of its flight hop— in other words, as much as 120 miles from base, and may indeed have been lost for 30 minutes or more. At the speed of an Avenger this could have put the flight 75 miles even further out in the Atlantic. This would have caused Fort Lauderdale to immediately send out a Ready Plane guarding 4805 kilocycles. The chance they would have brought back the flight would have been very good given immediate action, action that was dependent on simply knowing the actual ETA.

None of them certainly suspected a gaff during the drama of that night. But Donald Poole later must have suspected. In response to the Board's question "where did you estimate the flight to be" when it first signaled it was lost, he replied: "According to navigational data available, it was found, allowing 20 minutes for bombing over Hen and Chicken Islands, that Flight 19 should have been over landmark Great Sale Cay at the time of the first reported difficulty."

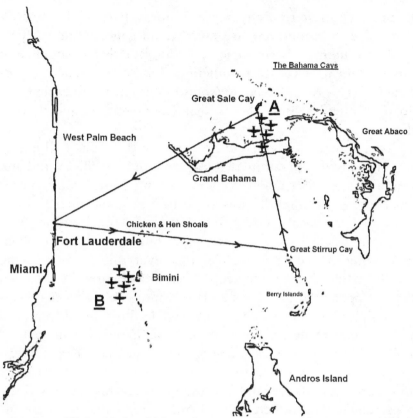

A, where Flight 19 should have been at 3:40 p.m., if on course, calculating from the correct time of takeoff, 2:08 p.m.; B, where Kenyon thought Flight 19 was, based on the erred ETA.

This is perfectly correct. His calculations are based on the only time he will ever refer to: the time of *takeoff.* Poole *never* mentions the ETA. Despite his voluminous answers putting in place minute time periods as to when he was called, when he went to Operations Radio, when he advised the Ready Plane to prepare, etc, Poole never mentions the flight's ETA. In fact, on his first day of testimony (he was recalled twice), his answer to "What did you do?" was 560 words, one massive long paragraph in the Board testimony that must have done nothing but confuse the Board.

The Board, however, should have gone over his testimony in more detail. By the time Poole had been called they already had Kenyon's testimony, which had placed Taylor around the Biminis, far from Great Sale Cay. Poole's testimony was openly in direct contrast to Kenyon's testimony. Part of the Board's eventual conclusions was to note that, in general and applying to all testimony, discrepancies existed. But this is simply not enough and it is far from judicious. Discrepancies in this instance were crucial. Kenyon's testimony had subtly led the Board in one direction, that is, that Taylor was completely derelict.

For instance, Kenyon confirmed that the "Yagi" [homing beacon] was in operation that day and that Flight 19 did not switch to the 3000 kilocycles emergency frequency— standard procedure— although he was asked to do so; and then Kenyon confirmed the weather was fine during the flight's time. This testimony could have no other effect but to established, albeit in a rather skewed perspective, that everything was working correctly, that they on the ground responded correctly, but that Taylor apparently did not. The Board, now believing that Taylor was close to Bimini, had to believe Taylor refused to ascend high enough to detect the homing beacon and that he refused to switch to the standard emergency frequency. Kenyon was the second person called to testify, and he was also the hands-on Operations Officer, and because of all this his answers clearly prejudiced and directed the Board in their line of questioning many of the succeeding witnesses.

This was apparent in the questioning of Lt. Richard W. Kohlmeier, the radar/radio maintenance officer. The gist of the questioning was essentially, if Flight 19 had been as close to Bimini as this ETA from Kenyon implied, then certainly their radio reception should have been perfect. Also, their IFF signals should have been picked up at base. Equally, they should have been able to pick up the Yagi homing beacon from Fort Lauderdale. Kohlmeier confirmed everything was working and that all the pilots had been instructed in emergency procedure. The Board had to already suspect there was something wrong with Taylor and his leadership. He must have been interfering with standard procedure to an unbelievable degree.

When Lt. Arthur Curtis, the Aviation Duty Officer, testified the Board's interest in the pilots was obvious. "Did you notice anything unusual about the behavior of the pilots in this flight?"

"Yes," he replied. "Lieutenant Taylor did ask me to find another instructor to take his place, giving me no particular reason except that he did not want to take the hop."

What another, less close-lipped officer might have elaborated upon we can only guess. But now the Board knew that Taylor didn't want to fly that day. The next question was obvious—the time he asked to be relieved. Curtis told them, "to the best of my knowledge," that it was 1:15 p.m.

"What was your action on this request of Lt. Taylor?"

"I said I would not be able to get him a relief."

"Do you know Lieutenant Taylor's movements from that time until the flight was dispatched?"

"I do not know, sir."

"Did you observe any unusual behavior of Lieutenant Taylor?"

"No, sir."

"Did he seem normal in all respects?"

"Yes, sir."

Curtis had previously confirmed that Taylor was actually late and that the flight took off about 15 minutes behind schedule as a result. It seemed now that this could mean Taylor had some other problem. The Board called one of Taylor's Torpedo 7 comrades, Jerome K. Rapp, as basically a character witness. "He was an above average pilot in the Fleet," Rapp testified. "He performed all his duties very well and was trusted and respected for the way in which he performed his duties."

The Board then followed their hunch:

"Do you know of any circumstances that might have contributed towards producing a recent mental strain or affected the physical well-being of Lieutenant Taylor?"

"I do not."

"Do you know of any recent change in Lieutenant Taylor's attitude toward flying?"

"I do not. Lieutenant Taylor liked his flying very well. He was very happy doing it."

By the second day, the Board already seemed to be homing in on Taylor, thanks largely to Kenyon's testimony. His testimony is probably basically innocent in its errors. Yet the image it painted of Taylor, albeit circumstantial and perhaps even inadvertent, was false. It was ultimately the Board's fault for not challenging any of it or dismissing it when debating their final decision.

By the third day they had more than enough reasons to be suspicious, too, about Kenyon's grasp on the whole situation. They now had the testimony of Lt. Samuel Hines, the Tower Duty Officer. As much as Kenyon, Hines was a central figure in the clique of officers at NAS Fort Lauderdale who set the pace of the whole rescue operation and in disseminating the information about Flight 19 to the rescue units. His answers really don't indicate there was any concerting of testimony between he and Kenyon, but they do suggest that the Tower officers were worried about dereliction of duty charges.

"What did you do?" prompted a long response from Hines that could not possibly allow both Hines' and Kenyon's "memories" to be accurate. Hines claims that at 4:06 p.m. Operations called down on the squawk box to his duty desk and said "Lt. Hines, please come up to the Operations radio. We believe we have some planes lost." Hines said he "ran up to the Operations radio and asked for the story so far from the radioman."

Hines paints himself at the double to get on the phone a few minutes later and call Port Everglades rescue facility (Thompson). To his frustration the line is busy. He then went back downstairs to discover that the line was engaged by Kenyon, who "told me he was advising them. So I immediately went back to Operations radio at which time a message was coming over, 'Does Lauderdale have any radar that could pick us up?' I immediately called the radar shack and asked if they had any equipment that could pick up emergency signals from a plane. The answer was negative and this was transmitted to the planes in distress."

In Kenyon's statement, however, he said that he took the call, not Hines; that he thought it was not an emergency so he sent Hines up to Operations Radio to find out. It may be a small example of coercion, but Kenyon must have seen the Board's reaction when he said he didn't think it was a big deal. In some small way, the next day, Hines is now making Kenyon and himself look much more on-the-ball.

Yet it doesn't seem to fit. Hines' scenario doesn't allow for

Kenyon to even know what flight is lost or what the situation is. Yet when he comes down stairs (only 4 minutes later) Kenyon is portrayed as on the phone with Port Everglades filling them in on details it seems unlikely that he could have had.

Hines' testimony is probably a good reverse barometer of what Operations was really like. None of them really tried to figure things out quickly, no doubt because of that erred ETA. From their vantage point there truly was no reason for them to think there was anything urgent about the whole incident so far.

Now, however, there were reasons why officers would want to start making themselves look more alert. Within days after the search was over, and hence as the first witnesses were being called by the Board of Inquiry, Robert Cox was making no bones about his view that there was "dereliction of duty" on the part of the ground operations at Fort Lauderdale. Specifically, he named Poole because of his refusal to send out the Ready Plane. With the entire flight now declared lost, Fort Lauderdale was faced with a possible national scandal.

When Cox was called and prompted to speak about what happened during his flight, the only indication he gave of the controversy now brewing was in his statement about his request to take out the Ready Plane. "I went to Operations and told the Operations Duty Officer (Hines) as much as I could about it at the time and I also asked him for permission to take the ready plane out. I didn't get any definite answer so I went up to the tower and Lt. Commander Poole, the Flight Officer, was up there. I asked him about taking the ready plane out and he very definitely said no, that he didn't think there was any use in sending it out then."

No man who holds the Navy Cross does so by a mere mistake or sudden impulse of action. Cox was brave, calculating, and instinctively intelligent. No one had to pry information out of him. He offered it readily, including one very correct deduction. "After, I came in and landed, with the intention of getting the ready plane and trying to go out and find the missing planes. I thought I would fly up in the northeast section and see if I could pick up the transmissions."

The radio fix had proven that Cox was instinctively correct

in his deduction. The extent to which he must have made an issue of it in the ensuing days before the Board of Inquiry was held is obliquely detectable in Poole's testimony *before* Cox had even been called. When Poole first testified, he made a clean breast of everything he did (without even being asked), bringing up and qualifying why he didn't send the Ready Plane. Poole's eagerness makes it plain that the controversy was an established issue; and, of course, it was becoming an even bigger issue now that the search had been discontinued and all were declared lost.

In every deduction Cox had turned out to be correct. This possibly sharpened his determination not to relent about his negative opinions of Fort Lauderdale's staff's competency, especially Don Poole's, though officially he didn't mention anything in testimony. Also, the fact he offered to take the Ready Plane out, though that was not his job, might mutely suggest he had experienced a lot of deficits in Fort Lauderdale's staff during his stay there.

Donald J. Poole was very diminutive, as many of the pilots were. The air forces like smaller, more agile men. Poole fit that bill perfectly. He was small, skinny, and although he would not show it to the Board he could have as short fuse. Unfortunately, in his first day of testimony he proved he had a long tongue.

Concerning his certainty that he knew Taylor could not have been over the Keys: "I then learned that his first transmission revealing that he was lost had occurred at 4 o'clock. I knew by this that Lt. Taylor could not possibly have gone on more than one leg of his navigation problem and still get back to the Keys by 4 p.m. Since Port Everglades was in touch with the flight, I immediately notified them to instruct FT-28 to fly 270°, also to fly toward the sun. I know this was transmitted because I listened over Operations radio. I do not know that it was ever acknowledged."

Poole's tongue had thus wagged toward making a clean breast of this other controversial issue. Was Flight 19 indeed told to fly west? Poole certainly made a point of accentuating he had heard his order to fly west broadcast to the flight, although the Board hadn't asked him. As time went by, to the Board's credit, they seemed suspicious of this last statement. By the time of Poole's testimony they had had the Fort Lauderdale radio logs (given them by Hines) and they could not find any record of Poole's instructions to fly west.

Lt. Azariah. G. Thompson, the officer in charge of Port Everglades, was the logical one to ask. Aside from Bammerlin and Cox, Thompson stands out as having tried to do an excellent job that night in assessing the situation. It is unfortunate in this light then that he rendered what is the lamest reply given by any witness before the Board.

—"We have testimony that Nan How Able 3 [Port Everglades, Thompson] was directed by Lt. Commander Poole, Naval Air Station, Fort Lauderdale, to direct the lost aircraft to fly west into the sun. This transmission was heard by the Naval Air Station, Fort Lauderdale radio [Poole's claim], but does not appear in Nan How Able 3's log. Can you confirm this transmission?"

And what was his answer? He looked at his log. "Yes, I can as follows: at 5:15 p.m. the following transmission was received from FT-28, 'I am flying 270 degrees. We will fly that until we hit the beach or run out of gas.' We gave them a 'Roger' to continue to fly that course."

To even assert this even closely bears out Poole's claim is ludicrous. To say that a mere "roger" is to be equated with an order to fly west is a statement of absurdity. "Roger" merely means an acknowledgment of reception of the message. It does not carry with it a command or agreement. The fact is the flight was *never told* to head west. Poole claims he gave the order at 4:45 p.m. Only Poole seems to have heard this over Fort Lauderdale's Operations Radio. Thompson's testimony mutely confirms it must never have been sent.

From the air-to-tower dialog we know that already at 4:31 p.m. Taylor had said: "One of the planes in flight thinks if we head west we will hit the beach." This time period is even before Poole claims he ordered the radio operator to send the order to fly west. The fact is that the only evidence we have about direction is Taylor asking a question; and nobody replied coherently to him about which direction to head. Amazingly, the radio operator was never called to testify. He could have elaborated on the lines spoken by Taylor, since he was jotting them down, but for this question he isn't necessary. The logs speak for themselves. Taylor was *never* told to fly west.

There are indeed too many lapses of time in the logs and memories of the officers at Fort Lauderdale in the first hour, but by 5 p.m. there is too much data and all the logs essentially jive with one another. It is clear by this time that Flight 19 was heading out into the Atlantic on a northeast course.

Thompson and ASR's insistence on sending the Dinner Key dumbo northeast over the Bahamas was a triumph of analyzing and assessing the situation. In comparison, Poole had not been assessing it well. The logs show that Thompson was certain the flight was lost northeast over the Atlantic. Poole, on the other hand, had only testified he was ready to send the Ready Plane out on a "direct easterly course" at 5 p.m. That would probably have accomplished nothing. The Ready Plane would merely have been heading straight east, far too south of Flight 19's course to pick anything up.

Poole may not have had a good hold on the situation, but he had done his homework afterward. His assessing of Flight 19's position at Great Sale Cay (if on course) was accurate and should have impressed upon the Board, had they been alert, that Taylor was not acting negligent in not detecting the base's Yagi beam. At this point, 120 miles from base, he was too far out. However, in going over the navigational data it would have been *impossible* for Poole *not* to have uncovered the ETA mistake. This is perhaps why he never refers to any ETA in his entire testimony. Would he offer that information? When the Board was finished with him, and he was then told he was privileged to offer *any relevant facts* which had not come out in questioning, he in fact did offer to speak, one of the few that ever did. . .but not about the ETA error.

"No sir, I have no facts, but I would like to convey the following information. Naval Air Station, Fort Lauderdale's last pilot fatality in the VTB training occurred 9 December 1944, all with a total of VTB flight training hours at 78,627.6. This is up to and including 4 December 1945."

Translation: our record was pretty clean up to December 5, 1945.

He was dismissed.

Poole's first recall (December 13) proves that the Board did not catch the gross difference between his and Kenyon's estimates of Flight 19's position. During this session the Board dwelt upon the ZBX equipment first. The apparent inability of the flight to use this standard homing device was a puzzle only because the Board was still stuck on Kenyon's feckless surmising that the flight was just east of the Biminis when it was lost, a mere 60 miles from Miami and Fort Lauderdale.

By the time of his 3rd recall (December 27), Poole's actions show that he had sensed that the Yagi homing beacon strength might be followed up by the Board. Sure enough, he was asked plainly by the Board if he had ever directed a test flight "to be conducted to determine the range" of the homing transmitter from the NAS. "Yes sir, I have." He did, in fact, not so coincidently perhaps, order it on December 14, the day after his last recall before the Board. The answer is quite surprising:

"The result of the test conducted by Lieutenant Jackson . . .are as follows: At 1000 feet, YG transmitter signals were received for a distance of 42 miles [not even as far as Bimini]. At 2300 feet, signals were received 56 miles. At 5000 feet, signals were received 80 miles. At 7100 feet, signals were received 100 miles. At 8200 feet, signals were received 115 miles. At 10,000 feet signals were received 123 miles."

It cannot be coincidental that Jackson ascended altitude in the same increments as Taylor had ascended— 2,300 and then 4,500 feet. Taylor obviously had been following procedure by trying to locate the beam. He never did. This should have told the Board right now he was lost further out than Bimini, and Kenyon's estimate was therefore worthless.

For the sake of clarity, Poole should have confessed that the test revealed the flight was clearly lost far beyond range. Taylor was obviously not willfully violating standard procedure. More than any other officer, Poole knew this. Timewise he had already calculated that Flight 19 could have reached its furthest landmark— Great Sale Cay— when the first messages about being lost had been picked up. Flight 19 having gotten lost this far out, Poole had to have suspected that the "broken land" which Taylor reported must have been the Bahama Cays north of Great Sale Cay, themselves over 150 miles from base. No matter how high Taylor ascended here there was little chance he could have detected the homing beacon.

It is obvious that the testifying officers did not elaborate except to exculpate themselves. But the Board should not be excused for not noticing some of these gaffs. In setting down the circumstances it was obligated to set down circumstances in context, especially when they rendered 56 Facts and 56 Opinions based on those facts.

Not only did the Board not dismiss Kenyon's early estimate, it continued to labor under his implication that Taylor willfully did not abide by standard procedure and switch to 3000 kilocycles. Sadly, this remained true even after the radio logs were introduced by Hines and Thompson. Within the dialog it is clear to see that Dinner Key told Taylor to remain on 4805 kilocycles so they could try and pick him up for the DF Net. By the time Port Everglades is almost pleading with him to change channels so that all Florida could pick him up it was already dark. Taylor's refusal was not reflective of belligerence. There was a 12 foot cable that connected the dash tuning apparatus to the radio equipment under and behind the pilot's seat. Over time "play" developed in the tuning button. When this happened the pilot had to really play with the knob to get the right channel. In the dark there was a very real chance that Taylor might lose contact with a pilot. He said as much when he replied: "I must keep my planes intact."

Despite rendering Facts and Opinions, the Board never challenged Jenkins testimony in another key matter. In his answer to Question 9 "When you first were advised of the flight's difficulty [at 4:30 p.m.], what remaining time in the air did their fuel supply give them?" Jenkins told them "I estimated they would be out of gas in about three to three and a half hours, or that they would not remain in the air after 8 p.m."

If so, this does not explain where the many inaccurate fuel estimates had come from. Several stations had been operating under the instructions from Fort Lauderdale that the flight had fuel until 6:30 p.m., then 7 p.m., then finally the 8 p.m. estimate which Jenkins is sticking to here. In fact, ASR informed the RAF at Windsor Field in their December 6th summary that the flight was one and a half hours long and "Had five hours of fuel aboard." There wasn't one estimate that was the same until Jenkins testified to this 8 p.m. time above.

Jenkins does not offer, nor was he asked by the Board, to go into the procedure for "throttling back" an Avenger and therewith extending its fuel endurance. Although it would be traveling at a slower speed, it could remain in the air much longer. Any Avenger pilot, and every member of the Board and especially Jenkins, knew the procedure for changing the mixture of the fuel. *Solomons'* radar report of those planes at only 120 mph indicates that Flight 19 did indeed throttle back.

The fact is Flight 19 could easily have been up past 9 p.m. that night. Every Avenger pilot I've spoken with is surprised by a 5 ½ to 6 hour estimate. Bill "Smitty" Smith even sent me extracts from an official journal on Avengers, highlighting the fact there are two separate calculations for combat and patrol range. Herein is found Jenkins rather clumsy mistake. He was estimating the "combat" weight and performance for an Avenger which includes the weight of a torpedo and accelerating to attack speed (285 mph), which gobbles loads of fuel. Its 1,200 mile range and 8 hours flight endurance is cut down to about 7 because of these factors.

The patrol range estimate is entirely different— up to 1,450 miles and 10 hours flight endurance— because there is no great weight of a torpedo to be carried during patrol, just the men and the fuel. This is the condition that the planes of Flight 19 were in. They never went into attack speed (Bimini was a glide bombing practice, not torpedo practice). Jenkins (and later, Poole) confirmed the planes were topped off, so there is no question they were fully loaded with fuel.[13] With this estimate, Flight 19 could easily have been 300 or more miles away from the 5:50 p.m. position fix and still flying at 9 p.m., an hour later than he estimated. What was Jenkins thinking?

The Board did not pull that thorn. They should have gone over all the testimony in order to sift out the puzzling and aggravating contradictions they had. But in their own rather doctrinaire estimation of faithfulness, they had served their purpose and fulfilled their orders. They "set in order the circumstances."

The real circumstances were: Fort Lauderdale didn't know

---

13 Poole also personally confirmed this to me.

the ETA; they weren't sure about the fuel estimate; didn't transmit to the flight to fly west; didn't react quickly; didn't relay messages; didn't send the Ready Plane; knew Taylor was told to remain on 4805 kilocycles, etc. but didn't tell other stations to switch over.

Poole, like Kenyon, Hines, Jenkins, and others, must have personally realized there was too much that didn't fit. Cover-up is a strong word and indicates an awful lot of coercion. But there was at least the lie of omission in Poole's testimony and negligence in the Board's assessing of the blatant contradictions between the testimony of key officers.

There is no need to guess that every officer was afraid of blame. Newspaper reports were getting worse. On his second recall, Poole was faced with 4 hardened officers who wanted it straight from the horse's mouth about which officer was "actually in charge" of coordinating the search and rescue activities at Fort Lauderdale that afternoon and evening. Poole responded clearly. "I was in charge." It was his shortest answer.

The Board had reasons to recall Poole and ask such a question. Officers involved in Air-Sea-Rescue had now testified. Each basically exculpated themselves of any guilt by subtly blaming Fort Lauderdale. All their actions, they testified, essentially were in reaction to what Lauderdale was telling them. To some extent this was true, and the excuse was especially effective when it came to the question of whether the radio position fix was transmitted to the flight. Yet this controversy cannot be entirely blamed on Fort Lauderdale. The lack of initiative at ASR was extraordinary.

Without dragging this on for too many more pages, it is nonetheless necessary to explore the testimony of a few more officers to fully bring out the underlying theme that can be easily gleaned from their testimony. That theme is one of sheer and utter lack of initiative to analyze, act, investigate and respond, especially when those officers from ASR in Miami and Banana River were called. Some of the lack of initiative was directly the result of the false ETA and Kenyon's interpretations there from.

On the 8th day of testimony (December 19, 1945) this was poignantly revealed when Lt. Commander William T. Murphy was called. He was the assistant to the Assistant Operations Officer (Commander Richard Baxter) for Air Sea Rescue at headquarters in Miami.

Question 6 of the Board (asked by Lt. Cmdr. Roberts) was to the point: "At what time did your activity takeover control of the Air Sea Rescue operations for the missing TBMs from US Naval Air Station, Fort Lauderdale?"

"At 4:40 p.m. we received the message from Fort Lauderdale. We alerted the Eastern Sea Net and the Gulf Net, and we stood by waiting to get bearings on the Nets. At this time we thought it was like any of the common occurrences that we have quite frequently on planes lost or overdue and did not consider it an actual distress until approximately 7 pm." This is amazing in light of his next sentence. "We were advised around 5:30 p.m. by Fort Lauderdale that the planes had fuel for another hour and 15 minutes, and if they were only working off Bimini, which we were told over the phone, they should reach the mainland."

This still does not excuse ASR for not considering the flight in emergency until fuel starvation (according to this estimate in the roulette of Jenkins' ever-changing estimates). Moreover, the bearings obtained for 5:50 p.m. should have been enough to stimulate more immediate action on ASR's part, since this fix proved the flight was far out to sea. It alone is what finally made Dinner Key send out its dumbo at 6:20 p.m.

That must have been on the Board's mind as well:

"When did District Air Sea Rescue Headquarters receive its first fix of the missing planes on the DF Nets that had been alerted?"

"The first fix we were able to establish was sent out on the net at 6:34 p.m., and it gave an approximate fix of 29° 15 minutes north, and 79° 00 minutes west, a radius of 100 miles, by averaging all bearings and times, that would be his position at 5:50 p.m."

That wasn't the question, was it? When did they *receive* it, not when did they first transmit what they knew. The span between both is actually over 40 minutes.

"What use was made of the position established by this fix?"

"It was immediately telephoned by Lt. (jg) Sorenson to

NAS, Fort Lauderdale who were in communication with the planes. That went right out on the Eastern Sea Frontier net (teletype net), which Banana River and WAX (which is at Hialeah) received. It was also sent out on the Gulf Sea Frontier HF/DF Nets which includes Dinner Key."

The Board recessed for lunch and during their meal they were obviously piqued by this inefficiency. They might also have recalled that Rolland Koch, the Tower Controller at Fort Lauderdale, testified he called the DuPont building (where ASR HQ was located) around 4:15 p.m. Murphy now claims it was 4:40 p.m., then again at 5:30 p.m.

When the Board members returned at 1:30 p.m., Murphy was recalled.

"Who was in actual control of Air Sea Rescue operations at Joint Operations center in the DuPont building, Miami, when word was first received of the five TBMs being a difficulty?"

"I was, sir."

"How long did you continue in that capacity?"

"I was in charge of Operations until the arrival of Commander Baxter at approximately 6:30 p.m. Commodore Benson, Chief of Staff to ComSeven, and ComGulf, was present on the bridge throughout this period."

Question 13 was aimed at the bull's eye. "What is the procedure in the Air Sea Rescue organization for transmitting positions obtained by radio direction finders to vessels or aircraft in difficulty?"

"The fix is telephoned or transmitted to the station that is in contact with the planes or vessels and he, in turn, tries to get it out to them."

"What station did you consider in contact with the aircraft and did you transmit the fix to that station?"

"Lt. Sorenson transmitted to fix via telephone to NAS, Fort Lauderdale."

"Does Air Sea Rescue Headquarters have any special radio transmitting apparatus available to them for transmitting direction finder positions to ships or aircraft?"

"Yes sir. They can go out through WAX at Hialeah or through Jupiter."

"Were either of the stations just mentioned directed to

transmit the radio direction finder fix to the missing planes?"

"Not to my knowledge, sir."

Murphy's testimony, though doing him no good in the eyes of the Board, directly and indirectly pointed the finger back at Fort Lauderdale. Directly by telling the Board that Fort Lauderdale had the fix; and indirectly, from our vantage point now, of showing that ASR's responses were sluggish because of the ignorance of the core group of officers at Lauderdale regarding the ETA and probable position of the flight when first lost.

Murphy's response was essentially echoed by the Gulf Sea Frontier Net officers, Coast Guard Commander John Harding and then Lt. Sorenson.[14] Sorenson especially testified to the fact the High Frequency bearings really were so poor there was no possible fix until 5:50 p.m. But why were no further fixes taken? That question was not asked by the Board. As regarding contacting the flight, Harding confirmed it would take only moments to switch over to 4805 kilocycles to listen, but it would take about 20 minutes to retune the radios in order to transmit. But this was never done. The Gulf Sea Frontier Net basically felt they did their job by transmitting the fix to ASR. ASR, as Murphy had testified, felt they did their job by telling the station they presumed in contact with the flight— Fort Lauderdale.

The Board recalled Lt. Commander Murphy again, this time for only one question. A very pointed: "What activity controlled all search operations for the five missing TBM aircraft under investigation?"

"Gulf Sea Frontier Air Sea Rescue Headquarters, Miami, Florida." In other words, himself, Commander Baxter and Commodore Benson. Remember, once again, they then blamed Fort Lauderdale.

After Poole admitted that he was in charge at Fort Lauderdale (during his recall), the Board brought up the telephone call from Port Everglades in which Thompson expressed his thought that the flight was lost over the northeast Bahamas and that a Ready Plane should be sent out to establish contact with

---

14 Sorenson basically "had control of the net which connects all high frequency direction finders . . ."

the flight on 4805 kilocycles. Poole must have thought this was a bad dream. It was the Ready Plane debacle again. Four ranking officers were glaring down on him looking for a fall guy. And indeed Poole was the last option they had left. Poole's answer was that he had not been directly told of this, "nor was I informed that it came from Air Sea Rescue Task Unit No. 4." He said rather that Kenyon had offered the idea to him only as a suggestion. When asked again to tell the Board what his decision was, he replied yet again with the masterpiece of illogic that he held the Ready Plane for fear that it might interfere with the tenuous communication they already had. (Hines had made it clear in testimony that they never had two-way communication with the flight.) Also, Poole said he wanted more definite information on where to send it.

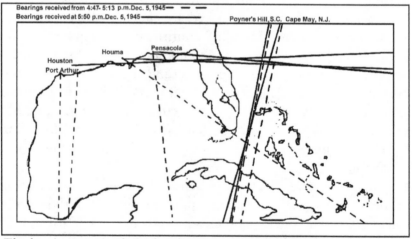

The bearings received. Broken lines represent early bearings. Solid lines indicate bearings used to plot the 5:50 PM fix. Note that some may actually have homed in on Cuban radio stations.

Poole was asked pointedly: "Did you know that Lieutenant Cox in FT-74 had been in radio contact with FT-28 at about 4 p.m. and that FT-28's signals had faded when FT-74 flew south?" Poole's answers were much shorter this session: "I did not know that until the following day." It seems amazing that Cox hung around the tower until 6 p.m. and that he would not mention this.

Now, even worse, they asked him a question on the radio

fix. Now that a definite location was known, at least definite enough to say with dogma they were in the Atlantic, a radio message to the flight would seem imperative.

"Was US Naval Air Station, Fort Lauderdale in communication with the TBMs when they received the position obtained by the fix?"

"No sir."

"Did you request Nan How Able Three, Air Sea Rescue Task Unit #4, to transmit the position obtained by the fix to the TBMs?"

"No sir, I did not."

If the Board was really looking for whom to blame, they must have realized that incompetence was rife throughout the whole search operation. As more and more officers and men were called, it became painfully obvious.

This was poignantly demonstrated by the slough of Banana River officers that were questioned in the next days. There were no controversies until Bammerlin's testimony on December 18, when he clarified their radios had never been set to guard 4805 kilocycles. At the last minute the Tower called down and had them switch to 3000 kilocycles since that was the usual channel covered. Apparently, Banana River never understood or could never appreciate the fact that Flight 19 was on 4805 kilocycles. It was therefore impossible for the PBMs to have ever picked up Flight 19's messages, although both the PBMs and Flight 19 were up at the same time (at least officially until 8 p.m.)

Then there is the question of why Banana River never picked up the flight on its radio. The 5:50 p.m. position fix shows at the very least the flight was closer there than Fort Lauderdale and Port Everglades, both stations still picking up bits and pieces of the flight's faint messages. Lt. Floyd Parmenter testified that he instructed the Banana River Tower to set up on 4805 kilocycles and contact the flight. "They told me that they did not guard this frequency and that it would take some time to transmit on that frequency. I told them to do the best they could and report any messages."

Thomas Tramberg, Seaman 1st, the Tower Controller, contradicted this in later testimony. "Mr. Parmenter didn't tell us to contact the missing TBMs, to my knowledge. I was there. He told us what information he had on the missing TBMs, their last known position, and asked if we could set up a listening guard on the frequency of 4805 kilocycles, and that is all we did."

In light of Harding's testimony, Parmenter's testimony seems more accurate than Tramberg's. Harding had testified that it would take time to tune the radio to send on that frequency, and Parmenter, too, testified that he was told that. This is specialist information a line officer would not necessarily have unless he did indeed ask the radio operator to transmit on that frequency and was given that exact answer in response, as he testified. Banana River might have had a chance to contact the flight had they set up to transmit. It would have taken them 15 to 20 minutes to do so. Flight 19 would still have been in the air.

But listening on 4805 kilocycles was just a turn of the knob away. Tramberg went on to state that they never heard anything on 4805 kilocycles. However, the radio fix truly calls into question whether Tramberg and Banana River were ever really listening at all.

Lt. Duane Walker, acting as the Duty Officer in the next office, was helping out Lt. Parmenter in the Air-Sea Rescue operations by fielding the long distance phone calls that were coming in from other stations. He, too, contradicted another Air Sea Rescue Officer's claim (Lt. Azariah Thompson). It is particularly damning testimony.

After examining the logs presented to them, the Board had asked a very lengthy and significant Question 6 to Thompson: "At 6:24 p.m. the Air Sea Rescue Task unit #4 [Port Everglades] communications log indicates that a message from Joint Operations Center, Miami, was intercepted by Air Sea Rescue Task unit #4 and forwarded by landline to Operations, Naval Air Station, Banana River. This message was to advise them to have all stations in the area to attempt to contact the flight leader, FT-28, on 4805 kilocycles and advise him to maintain the course of 270 degrees. Can you verify that this was accomplished?"

"I can," Thompson had replied. "I made the call myself by landline at about 6:24 p.m. In that conversation I told them to tell the planes to steer 270 degrees."

Walker would have been the one at Banana River who would have taken the call. When he was recalled to testify, he was asked pointedly in question 1: "Previous testimony states that the Air Sea Rescue Task Unit No. 4 [Thompson] informed Naval Air Station, Banana River, by phone to contact all stations in the area to work FT-28 on 4805 kilocycles and to advise them to maintain the course of 270°. Did you receive this message?"

"No sir."

The Board was surprised. "Referring to the Communication Log of Air Sea Rescue Task Unit No. 4 of 5 December 1945, which has been introduced in evidence to the Board, at 6:24 p.m. Air Sea Rescue Task Unit No. 4 relayed teletype message from Miami to Banana River concerning the fix of 29° north, 79° west, and also advised Banana River to please have all stations in this area attempt to contact FT-28 on 4805 kilocycles and advise them to maintain a course of 270 degrees. Did you receive this message?"

"No sir, I did not."

The Board, maintaining its official aplomb and reserve, was nevertheless disturbed. Their next question was to ask if he knew of any similar message that came in. Walker said that at 6:10 p.m. he got a call from Port Everglades telling them there were 5 missing TBFs out of Fort Lauderdale. He was relaying the message due to teletype trouble north of Port Everglades. (Parmenter had confirmed that the teletype line was not working south of Banana River.) Walker was told by Thompson to alert the stations in their area and dispatch the Daytona and Vero Beach dumbos and "put a guard on 4805 kilocycles."

That was incredible testimony. A broken teletype line is nobody's fault, but a telephone call relaying this information is critical. Now it seems it hadn't happened.

"Were you cognizant of all telephone calls received by Air Sea Rescue, Banana River, concerning the emergency on 5 December 1945?"

"Yes, I was cognizant of these calls. The long-distance telephone calls that came in with reference to the missing aircraft were at 6:10 p.m. from Port Everglades and approximately 8:30 p.m. from Miami, and that is all."

Had Banana River received the call or gotten the teletype message, it is very likely they would have had enough time to set up to transmit on 4805 kilocycles, and by doing so they would have contacted the flight, subsequently told them to fly west or at least told them what the radio fix was, thereby making up for Fort Lauderdale's botch of not transmitting this fix to the flight earlier at 6:15 p.m.

Another controversy: "You have previously testified that the Boat Facility at Port Everglades advised you by telephone that the high frequency direction finder fix had established an approximate position on the TBMs of 29° north, 79° west. When did you first receive information regarding the HF/DF fix on the missing planes?"

"I receive that information at the time *I phoned* by long-distance the Boat Facility Officer [Thompson] at Port Everglades at about 6:36 p.m."★

"Did you receive any other information from Port Everglades concerning the emergency at that time?"

"No sir, I did not."

Commander Lawrence, the Commanding Officer of Banana River's ASR, who had been called earlier and asked only routine questions, was now recalled for 2 quick questions, essentially to confirm the teletype machine was working but nothing was coming in. (Obviously, the line was dead just as Thompson and Parmenter had said.)

Parmenter and Walker's testimony jives, leaving no doubt that no phone call came in telling Banana River to call the planes on 4805 kilocycles, telling them the fix or telling them to tell the flight to fly west, nor to inform any other station to inform the flight.

All information, whether from Fort Lauderdale or ASR in Miami had stopped at Port Everglades. The teletype line was down north of Port Everglades, so that the messages being sent by Miami were going no further north than Port Everglades. . .and Thompson clearly botched the job of telling Banana River. . .or Banana River's two officers were lying, which is unlikely. Parmenter, of course, claims that he did contact the

---

★ Italics mine

Tower and asked them to contact the planes on 4805 kilocycles, but this again was contradicted, as we know, by Tramberg.

It would have been a different matter if this order had come over the command line or the telephone from Miami. *But it had not* regardless of what Thompson had claimed. He never phoned it in to make up for the broken teletype line from Miami to Banana River. Even this Board would later regard this as incredibly crucial.

Enough said. The key points of the search and rescue debacle have been brought out here sufficiently by now. Although the Board would interview many witness over the next several days, the key officers left were those who were to be recalled, like Jenkins and Poole. We have already touched upon Poole's recalls. Questions to Jenkins essentially surrounded who was to blame. Jenkins said Poole, though his junior, was in charge. For our case here, it is necessary to recall only one witness.

This witness is Lt. Commander William T. Murphy. He was recalled by the Board on Friday, December 21, and asked to "evaluate" a number of the sightings made during the search. One included the radar report of the USS *Solomons* of those 4 to 6 aircraft picked up off the coast around Flagler Beach at 7 p.m.

"It was evaluated as negative."

The Board could not have challenged the answer more diplomatically.

"Was the identity of these planes ever established?"

"No sir."

Then exactly what does "negative" imply? What sort of criterion is used to establish the evaluations? Proceeding further into Murphy's testimony the answer becomes clear— negative means 'we don't know, but nothing ever came of it anyway.'

What we do know now is that there was no real evaluation to try and determine what happened to Flight 19. More than anything this contributed to its ultimate disappearance. The Board set in order the circumstances surrounding the disappearance. In doing this they revealed there was very little initiative or inspiration behind the search and rescue.

Based on all the above testimony, what could the Board

conclude? It was not necessarily within their power or their directive to render a conclusion. But they had these basic points: Taylor took over the flight because he thought they were off course. He then believed that instead of this his compasses had been malfunctioning; the flight must have been within range of the Yagi homing transmission beam when they first got lost, if only Taylor had just followed procedure and climbed altitude enough; he did not switch to 3000 kilocycles, standard procedure. The blame was fixed. It was Taylor's fault for everything.

Opinion 44, the second to the last dealing with Flight 19 (as opposed to the PBM Mariner), reads in part "That the disappearance" of the flight "was caused by a temporary mental confusion resulting in faulty judgment on the part of the flight leader and instructor of Flight 19, Charles Carroll Taylor, Lieutenant, US Naval Reserve, in permitting himself to lose knowledge of his general position relative to the peninsula of Florida, and failure on his part to utilize the facilities available to him in his flight for orienting himself." Opinion 45: "That no offenses are known to have been committed, nor serious blame incurred by any known person or persons other than that which may be assigned to Lieutenant Charles Carroll Taylor, US Naval Reserve, flight instructor in charge of Flight 19 for his faulty judgment that contributed to the disappearance of . . ." Every man's name is listed, every plane's number is given; all vanished. The end of that. Taylor is to blame.

The last opinion (56) is a triumph of understatement.

> That the various logs and records submitted in evidence by all the participating activities are incomplete, compiled from direct logged communications, memory, notes, and memoranda, and are difficult to reconcile one with the other and with the testimony in some cases, but do represent the best records available.

This verdict was not a public one. The whole volume of testimony and documents, over 500 pages, was in the hands of the Board's superior, Admiral Davison. On January 24, he issued a letter to the Secretary of the Navy, with copies sent to his direct superior, Admiral F.D. Wagner at Pensacola, and other copies sent to NAS Fort Lauderdale's commanding officer, William O. Burch, and Commander

Eastern Sea Frontier, Banana River, just about all those who were sweating to know.

Davison essentially copied Opinion 41 of the Board when he set down the ancillary factors contributing to the loss of the flight:

The convening authority notes in Opinion 41 of the board of investigation that the factors contributing to the ineffectiveness of efforts to locate and recover the aircraft of Flight 19 while still airborne were due to excessive interference on the training frequency in use by the flight; failure of the flight to shift to the Air Sea Rescue frequency; failure of the Air Sea Rescue teletype during the critical period; the inability of the HF/DF nets to obtain early and reliable bearings due to the interference on the frequency in use by the flight; and lack of complete coordination, understanding, indoctrination, and training in Air Sea Rescue procedures on the part of principal participating units, both within themselves and in combined effort. However, it is considered that the primary reasons for the disappearance of Flight 19 were the confusion of the flight leader about his location; his failure to take into account strong winds which apparently carried him farther east than he realized; and his failure to utilize radio aides which were available to him.

It is considered that the coordinated search operations conducted were thorough, adequate and well-organized.

The convening authority is, by separate correspondence, recommending to the Commander Eastern Sea Frontier that he conduct frequent Air Sea Rescue drills utilizing such facilities of this Command as he deems desirable in order to promote complete understanding among all units which might become involved.

Subject to the foregoing remarks, the proceedings, findings and opinions of the Board of investigation in the attached case are approved.

There it was. Complete absolution. Even those damning bits the Board was able to myopically dissect from the testimony Davison only regarded as factors that contributed to the "ineffectiveness of efforts" to locate and recover the flight while still airborne. Does that not, in essence, mean that the flight would not have disappeared otherwise? Paragraph 2 above, short and simple, was absolution for this. Yet it is a total contradiction to the language above it where it read "lack of complete coordination." The Board had actually been more critical than Davison would now be.

To his credit, Admiral Wagner, the overall commander, at NAS Pensacola, saw the obvious paper shuffling contradiction of paragraph 1 and 2. He would be a little more critical in his short letter to the Judge Advocate General. "The record of proceedings of the board of investigation in the instant case discloses several discrepancies in the execution of the training program which, in themselves, were of no vital import, but the sum total of which contributed to the loss of the TBMs. The convening authority's comment is concurred with that, in essence, the leader of the flight became so hopelessly confused as to have suffered something akin to a mental aberration."

Although the press and public were crying for information, the conclusion of the Board would remain private for some months to come. During that time, the Navy had to fight another battle.

1. Avengers in flight. *Navy History Center.*

2. Naval Air Station Fort Lauderdale, 1945. *National Archives.*

3. The Avenger was the largest single engine aircraft in World War II.

National Archives.

4. Avengers in formation over the field. *National Archives.*

5. NAS Fort Lauderdale circa 1960, serving as Hollywood/Fort Lauderdale Airport. The hangar, Administration, Torpedo Training and Ordnance Training buildings are in center.

Courtesy Alan McElhiney

6. William O. Burch, as a Rear Admiral.

Navy History Center

7. Don Poole, 1945.

Courtesy Don Poole

8. Charles Taylor's induction photos.
Navy History Center

9. Edward J. Powers,
Princeton grad. photo, 1941.
Pilot of FT-36
Courtesy of Susan Spengler

10.   George William Stivers.
Annapolis grad. photo.
Pilot of FT-117
Courtesy of Phyllis Anne Barry

11. Joseph Bossi, pilot FT-3
*Cherokee Strip Land
Rush Museum*

12. Burt E. Baluk Jr.
Courtesy Charles L. Baluk Jr.

13. "Robert Francis Harmon," the spunky
George Devlin Jr.
Courtesy Bill & Rita Smith.

14. Herman Thelander
Courtesy Fern Thelander.

15. Robert Francis Gallivan,
Induction photo
Author's Collection

16. Robert Peter Greubel
Induction photo
Author's Collection

17. George Paonessa
Induction photo
Author's Collection

18. Howell Thompson
Induction photo
Author's Collection

19. Billy Lightfoot,
    Induction photo
        Author's Collection

20. Robert Ford Cox, on the wing of
an Avenger.

NASFLHA

21. Junior Bachelor Officers' Quarters and, in foreground, the movie
    theater. Circa 1960s.

NASFLHA

22. Graduating Class 16, Fort Lauderdale, 1944. Bill Smith is back row 4th from left. "Bob" and "Rich" are below him.

Courtesy Bill & Rita Smith

23. Weldon Richmond and Bob Harmon at Coney Island, circa 1943/44.
Courtesy Weldon Richmond

24. Bill "Smitty" Smith
Courtesy Bill & Rita Smith

25. C.C. Taylor's last graduating class, Miami, August 1945.
Taylor is center.

Courtesy Barbara Gonzalez

26. Officers of Torpedo 7, South Pacific 1944. Taylor is 1st row, 2nd from right.

Navy History Center

27. Powers instructing ground Marines in combat tactics, Quantico,
Virginia, circa 1943.   *Courtesy Susan Spengler*

28. Captain Ed Powers, right, at a cocktail reception.

Courtesy Susan Spengler

29. Newspaper clipping of Michael Belvito's strange disappearance. 30. C.C. Taylor at a party.   (30 *Courtesy of Carol Ann Rossi*)

31. Stoic, Taylor squints at camera as the ship picks him and Nafstad up off Guam, 1945.   *Courtesy of Barbara Gonzalez*

32. Military bigwigs Julian Smith and A.A. Vandegrift in foreground. Stivers marked himself and the others. He was attached to Smith's staff at the time.

Courtesy of Phyllis Anne **Barry**

33. Stivers enjoying a cocktail party with some buddies. He was a very popular officer.

Courtesy of Phyllis Anne Barry

34. Stivers at a Miami cocktail party.
Courtesy of Phyllis Anne Barry

35. George Paonessa with his
    father, Frank Sr.
    Courtesy Frank Paonessa Jr.

36. George Stivers.
    Courtesy of Phyllis Anne Barry

37. Inside Fort Lauderdale's Tower.

38. The ZBX (inside Tower)

39. Operations Radio's gear.

40. Communications Room.

(photos courtesy NASFLHA)

41. Panoramic shot of NAS Fort Lauderdale and its sprawling airfield.

NARA

Chapter 13

"An Untold Agony of Suspense"

WHILE THE CLOSED-DOOR INQUIRY WAS ONGOING, THE PUB-
lic image of Flight 19's drama was harder, if not impossible, for
the Navy to control. The AP and UP wire services were tap-
ping out piquing messages, causing almost every paper, including the New York *Times,* to show interest.

More than any other anxious family members, Joan Powers,
Captain Powers' wife, believed that something unusual had
happened. By the morning of the 6th of December she was fully prepared for bad news. She had awakened in a jolting panic
at 2 a.m. in their Mount Vernon, New York, apartment. She
had never experienced this before. She sensed something must
be wrong, and called the base. All she was told was that Captain
Powers was not there, without any definite clarification why.
That alone was enough to worry her at 2 a.m. Now this morning she heard on the radio that five planes were missing from
the base. Until the official telegram and phone call arrived, she
would remain in suspense.

Within a day all of the families had been informed and were
desperately awaiting confirmation that the men had been found.
Instead depressing news would be posted on December 10. All

papers carried the story of the "last ditch search effort" and the gloomy summaries.

The wartime buddy relationship between George Devlin (Bob Harmon), Weldon Richman and their commanding officer-pilot Bill Smith continued to survive in the two remaining. On January 31, 1946, Weldon Richman wrote to his old friend "Smitty," now living in Peoria, Illinois.

Dear Smitty,

I suppose when you receive this you'll say, "Well it's about time." I'm really sorry about not writing sooner, it seemed like every time I'd start a letter to you someone would interrupt me. I hope I make it this time.

I suppose you are a happy civilian now. I've often wondered if you ever got out and what you're doing now. I hope you'll write soon and let me know how things are.

No doubt you heard the bad news about Harmon. In case you didn't, I'll tell you about it. He was in one of those five TBF's that are missing from Fort Lauderdale since December 5. I didn't hear about it for about 10 days after the accident. I certainly felt bad about it and his folks really took it hard. They still don't have any word on what happened, but have declared them dead. I hope that you heard about it before, because I know you'll not forgive me for letting it go this long without letting you know. . .

I have a leave coming up on the 12th of February. I plan on going to Brooklyn and see Harmon's folks. I've written them a few times and they want me to come. I'll be getting discharged in March and will be going home for good then, so I think I should go to Brooklyn while I'm still here in the East and have a chance.[15]

At the same time Weldon's letter reached Bill, so did a letter with a return address in Brooklyn— from George Devlin Sr. "Never having met you personally, it feels strange to write to you under these circumstances. However, feeling that you would want to know I am dropping this line to let you know

---

15 Weldon Richmond lived in Utah.

that George my son (Robert F. Harmon) is missing in a plane since Dec. 5th of last year." Mr. Devlin was not certain he had the right address, so he said he would "keep it short." He ends: "If you will drop me a line and let me know if this is addressed correctly, I will send more details to you about what has happened."

Bill Smith immediately wrote him back, receiving in reply a very significant letter. I think it is best to quote in its entirety this long letter. It illustrates so much of what was going on at this point in the drama of Flight 19; and what Mr. Devlin has to say fills in chronologically many of the events and much needed information that we need here, from all the families' vantage points.

The letter is dated February 10, 1946. It is eight 8x10 pages in a fine handwritten pen. The reader will be surprised at what the Navy was saying to the families, even at this date 2 full months after-the-fact.

Dear Mr. Smith:

I received your welcomed letter and was glad to know that I had contacted you because I know that you ~~were~~ was a good friend of Georgie's. He often spoke of you and he thought the world of you, to him you was a great guy. About Richman, yes we notified him. He expects to be discharged from the service in March, however he is getting a ten (10) day leave starting Feb 12th and he is coming to Brooklyn and he is going to spend his leave with us, and we are glad to have him because as you probably know, he used to come home with Georgie every couple of weeks when you were all stationed at Quonset Point and Squantum; and of course we had some good times together. I would have notified you sooner but I didn't get your address until after Georgie's personal effects arrived home. It was amongst these that I found a letter to him from you while you were still at Kingsville, Texas. There was two (2) addresses and I took the chance of writing to the La Veille St. address which proved to be right for which I am glad.

Well about Georgie, after he left Kingsville and arrived at Fort Lauderdale, he was very discontented. He tried hard to get assigned to a squadron that was going out to the fleet, but he had no luck. The flight executive officer proved to be a good Joe though, and he did all he could to make the boys happy by giving them good duties and otherwise doing all he could for them. George stopped flying about Oct, it was about that time that he ceased to draw flight pay, and of course this didn't make him feel any better. However, it suited us because we felt that he was better off on the ground and that he was comparatively safe. He had been assigned as an instructor in turret firing at the Beach Range, which for him was easy work. Of course with all this knowledge we were more than surprised when we were informed that he was missing in a plane. He expected a leave in Oct, but it was canceled. When he didn't get home for Thanksgiving, we felt certain that he would be home for Christmas and such would have been the case. In his last letter to us he told us that he was getting a ten (10) day leave starting Dec 22nd to January 3rd which would have been perfect, allowing him to spend Christmas with us, but such was not to be.

For on Dec 6th we were notified that George was missing with the flight of five (5) TBFs over the Atlantic Ocean. It was published in all the newspapers and we listened to the radio which was reporting the progress of the search for six (6) days. It was the most intensive peacetime search that had ever been undertaken, because this was the first time in Naval Aviation History that such a thing had happened.

We were kept informed by the Navy by telegrams of what was what. After the final telegram telling us that the search had been abandoned with— as they put it, no hope of recovery remaining, they sent us a letter telling us what had happened. The search went on for six (6) days and nights, with over 250 planes taking part, along with an aircraft carrier and numerous smaller surface craft, Coast Guard cutters, and amphibious jeeps, and so on. So it seems that they did all in their power to find these men. Here is what they told us in the letter.

A flight of five (5) planes left the station at 2:10 p.m. Dec 5th on a navigational training flight. They were due back at 5:23 p.m. and they were supposed to have enough gas to last them until 7^{30} p.m. At 4PM one of the planes radioed to the station that they were lost. The station radioed to them to return at once and what course to follow. The station received no reply to this message nor to any subsequent messages that they sent to the planes. At 5:23 when the planes failed to return, the Air and Sea Rescue activities were alerted and the search started. There were fourteen (14) men on these planes. A Martin Patrol Bomber was sent out from the Banana River Naval Air Station with a crew of thirteen (13) men aboard to give aid in the search. This plane either crashed or also was lost making a total of twenty-seven (27) men who were lost. So you can see that things were quite a mess at the Station. Services were held in the chapel at the station on Dec 13th for the men. They had a priest and the station chaplain Jack Courage, whom you might know, officiating at the services. A large American flag, (fourteen (14) of them) were used to represent the missing fliers. The chaplain sent us the flag, also photographs of the services. The Capt. Burch Jr. and the chaplain sent us lovely letters of sympathy, also the offer of their services in any way that they might be able to help us in our hour of sorrow. We also received a nice letter from Vice Admiral M.A. Mitscher who is now in Wash. as Deputy Chief of Naval Operations (Air). Of course this was all done in an effort to comfort us, in fact they did all they could for us, and we appreciated it.

George had notified the Navy of his having enlisted under the name of Harmon. We had sent certain documents to him, that the Navy had requested and so this matter was being taken care of at the time of his disappearance. The final step in this matter is in the hands of the Bureau of Naval Personnel, and what the outcome will be I don't know at the present time.

There is a party up in Northampton, Mass by the name of Gallivan whose only son was lost on the flight and who intends to find out the truth about this affair if such a thing is possible. His boy was a sergeant in the Marine Corps and he was due to be discharged Feb 20th. Well, his father obtained the names and addresses of the fliers with their next of kin, and he has started corresponding with all of us in an effort to see if we can get to the bottom of this thing or else help to prevent the loss of other boy's lives in the future. He feels that someone is very much at fault. So far he has found out some information. The leader of the flight was Lieutenant Taylors [sic], a combat pilot of two (2) years experience in the South Pacific. According to one boy's father, a sailor friend of his son, who was in Florida at the time of the tragic flight told him this story. There were two (2) flights of five (5) bombers each, and they were to fly in opposite directions, meeting over a lighthouse at Bermuda and then return. When they started out the weather was moderate, but during the flight a strong wind came up and blew them off their course. After radioing to the station saying that they were lost, one of the flights got on the beam and returned safely to the Station. They are supposed to have reported that after getting on the beam and starting back to the station they heard the other flight (with George) trying to ascertain where they were and which way to go. He wants to know why they didn't do something as long as they were on a beam, to help our boys in, instead of just thinking of their own necks and leaving our boys out there to their fate. If this report is true than the other flight must have been commanded by someone lacking in American Navy Spirit and who at heart is very selfish and cowardly.

Mr. Gallivan, after pestering Captain Burch for more information, finally received another letter, explaining just how the Navy figures this mishap occurred. A copy of the letter was sent to me.

He says that "The flight was led by a Senior Naval Aviator, with a student flyer handling the controls at the start of the flight under the supervision of the instructor. While off the Florida coast they passed over a group of islands, and for some unknown reason the instructor mistook these islands for the Keys at the southern tip of Florida. After becoming initially confused, it is further believed that the instructor then became uncertain as to whether he was in the Gulf of Mexico or the Atlantic Ocean, and in attempting to lead the flight back to Florida he flew in the wrong direction.

There is absolutely no indication that there was any malfunction of the airplanes or their engines, and they are considered to have been in excellent operating condition.

The weather on the afternoon of 5 December was average although the sea was rough and it is believed that the airplanes landed in the water late in the evening under these conditions. The aircraft were all equipped with life rafts, parachutes and survival equipment, but it is believed that in making a water landing after dark in the rough seas, it would be very difficult to use this equipment."

So Mr. Smith, there you have the whole story as we know it, for what it is worth. Of course a lot of it is nothing but baloney. It appears to me that there was a lack of coordination on the part of the shore facilities, knowing that there was a flight at sea, and later knowing that it was lost, it seems to me that they were rather lax or a bit tardy in doing something before it was too late to aid these lost fliers. Perhaps some of the 'Brass' or 'Braid' are being covered up, a thing we may never be able to prove, excepting of course if some of our boys should survive and so be able to refute some of the lies we believe are being told to us. After all, the Army and the Navy will never let anyone know anything but what they want them to know. And when you are up against a stone wall there isn't much you can do about it but hope for the best.

I believe I have given you this story as fully complete as we have it up to the present time and I haven't much more to say just now, however I will keep you informed of any further developments in the progress of this case, because being such a good friend to Georgie I know that you are interested in the outcome.

I wish to thank you for your response to my letter, also for your kind offer to help us. No one can help much just now except through prayers for Georgie and his buddies, that they may be found or else if they are lost forever then God's Will be done and may their Souls Rest in Peace.

So for the time being, and again thanking you for your having befriended Georgie, and your interest in him, also wishing you the best of health and luck in all your endeavors

I am

Sincerely Yours
George F. Devlin.

Fort Lauderdale NAS was seriously that blind to the actual circumstances that they continued to promote the nonsense that Taylor was merely lost 50 miles off the coast at Bimini, "the group of islands" that he supposedly misidentified. They even inscribed the erroneous ETA, presenting it on letterhead as full testimony of their careless investigation. The story about Bermuda is clearly a corruption of the Bimini story, and the two flights must be referring to Flight 18 having been on the same route ahead of Flight 19 but which came in without any difficulty.*

Mr. Devlin's letter touched on some behind-the-scenes developments that were significant— family and friends were beginning to contact each other. The Navy's silence in the face of such an "unprecedented event," as one naval officer called it, was a foolish blunder. Animosity and distrust were beginning to brew.

John Gallivan was a good Massachusetts Irishman, and he didn't feel a flag and letters of condolences were enough for the loss of a son. In his letters to other family members he expressed his belief that 5 big warplanes just could not be blown out to sea and disappear. Neither did Joan Powers. His letters struck a chord with her more than others perhaps because of her December 6 nightmare. She didn't know where the flight ended up, but she was convinced they did not go down at sea. Between them they started bringing in officialdom. Gallivan was writing Bostonian officials, and Joan and her father-in-law were lobbying their congressman.

Without the personal direction of Gallivan or political connections of Joan Powers, the Parpart family opted for something that they thought would prove more valuable to the "cause." They allowed the newspapers to publish some of the letters back and forth between the families, thereby bringing the nation into the private struggle of the families to find out what happened.

Charles Taylor's mother, Katherine, was beside herself at her son's disappearance. She wrote to the Parparts a letter beginning with "I have thought so much about you and all the other parents whose darling sons were snatched from them on December 5. I am almost a mad woman and to go on to the end without ever writing to any of you is something I can no longer do. I have been reaching out in all directions seeking some

---

* To this day, it is a part of the lore of Flight 19 that another flight encountered something odd, broke up and landed at several bases.

light, some verification of such a stupendous, unbelievable catastrophe."

Billy Lightfoot's mother wrote a heart-wrenching letter
back. "I have already gone through this once. Another son was
reported missing in 1944 and I suffered an untold agony of suspense of a year only to learn that he and his crew had been
killed on the day they were reported missing. Five months later
I received this horrible news of my youngest son. I am a nervous wreck and can't help but be very bitter."

The nation's interest escalated when a belated discovery in
the Florida Everglades fanned the flames of suspicion that the
Navy had been all fouled up in its search. "News has come of
some guides finding evidence in the Everglades, a most isolated
section, of a group of campers," read the newspapers. Mrs. Parpart wisely shared some more letters with the papers, including
the *Kansas City Star*, the paper that served the hometowns of
both Stivers and Bossi. The *Star* confirmed that her letters were
based on a phone call from Joan Powers, which "divulged that
elemental scraps were left in the small ashy heaps where fires
had been. Bones of very little birds had evidently been eaten
and picked clean." Gallivan's letter of February 25, 1946, was
dynamite.

> Dear Mr. and Mrs. Parpart:
>
> I am writing to answer your most welcome letter I received
> yesterday. So far as the boys are concerned conditions remain the
> same, except this, and I think it is vital and important.
>
> Now you live in New York, so get all the last week's issues
> of the Daily News, and look at Jimmie Powers' column, the
> "Powerhouse." He is at Miami at present. This is what it said:
> "Guides coming in from the Everglades report human tracks
> deep in Glades, leading them and many others to believe that the
> missing aviators are living Tarzan lives in the jungle."
>
> I had been in telephone communication with Mrs. Powers,
> Captain Power's wife, at Mount Vernon, and she said that three
> weeks ago an Indian went to Captain Burch and told him about
> it but the Navy does nothing. Now if those guides saw large
> numbers of footprints it must be unusual to them in that place,
> otherwise why would they report them! I think some of our boys

are in the jungle and we all have got to do something to get them out.

I suggest we post some money, to be paid by all of us, offering a reward to any guide that goes in there and brings out our boys. It is very tough country, and in some places nobody has ever been in there. Of course they have got to bring out our boys, not somebody else's.

Please let me hear your opinion on it soon because we can't wait any longer. They may be starving or sick.

John F. Gallivan

Publishing that letter angered the Navy and Marine Corp. It wrenched the hearts of millions of American while at the same time it made the Navy look stupid and careless. But it was plain the Navy beef wasn't with the Parparts; it was with Gallivan. Through the Marine Corps they knew he had even involved the Mayor of Boston, James M. Curley, who subsequently interceded on his behalf. By March 10, the Marine Corps felt obliged to respond to Curley and allowed it to be published in the papers— fighting fire with fire. The entire front page of Section C of the *Kansas City Star*, which was publishing the family letters, was devoted to Flight 19. A special box on the page had its own headlines:

FINDINGS ARE ON FILE MARINE CORP SAYS

Hon. James M. Curley, Mayor,
Boston, Mass.

Dear Mr. Curley:

Receipt is acknowledged of your letter dated January 28, 1946 and enclosures of Captain W. O. Burch Jr., United States Navy, and Mr. John F. Gallivan, dated January 18, 1946.

This headquarters has received an advance copy of the record of proceedings of a Board of investigation convened at the U.S. N.A.S., Miami, Florida, December 10, 1945, for the purpose of inquiry into and reporting upon the circumstances attending the disappearance of five TBM-type aircraft at approximately 1900,

December 5, 1945, resulting in the missing status of Naval personnel, including Robert F. Gallivan, Sergeant, U.S.M.C.

A review of the record and findings of fact and opinion made by the Board confirms in detail all points of information contained in Captain Burch's letter to Mr. Gallivan.

Due to the initial difficulty and subsequent failure of communications with planes during the critical period of flight on December 5, 1945, it is unfortunate that answers to some questions regarding reasons why the leader of the flight became confused as to his direction must be left to speculation based on such information as is available.

It is felt, however, that the Board's report represents a thorough and searching study of the disaster and indicates that Captain Burch's statements are correct.

Mr. Gallivan's anxiety in the matter is fully appreciated. A copy of the complete record of the Board of investigation will be on file at the office of the Judge Advocate General, Navy Department, Washington D.C., and communications in regard to such other information as is desired may be addressed to that office

A. A. Vandegrift
General U.S.M.C., Commandant of the Marine Corps.

Admitting that the Judge Advocate General's office in Washington held the copy of the findings was the wrong move for the Marine Corps. Through mutual friendship between the Carney family (of Mount Vernon)[16] and the Judge Advocate General, Joan Powers was able to make an appointment with Admiral Jennings. She came with a couple of friends, one of them sister-in-law Lenora Robertson, who was a stenographer.

While waiting to be received, Joan Powers instigated a feint. While some of the clerical officers were distracted, Lenny and Joan were able to get behind the office door where the document was kept. They began hand-copying parts of the Findings of Facts and Opinions. One was dynamite— Opinion 41 in which it stated in echelon of importance the contributing fac-

16 Same family as actor Art Carney.

tors: ". . .and lack of complete coordination, understanding, indoctrination, and training in Air Sea Rescue procedures on the part of principal participating units, both within themselves and in combined effort." She was not happy that the Navy's culpability, even if only a weak allusion to it, was being covered up. Nor was the Navy happy when they suddenly walked in and found her in restricted territory copying a confidential document. The upshot was a 'well done bad and unfaithful servant' and out the door.

This incident only caused greater distrust. Joan Powers had only seen a very small part of the document. What else the voluminous report contained was unimaginable. But there no doubt could have been plenty more unsavory details regarding the Navy. This quickly got around to Gallivan, goosing him to push the other families about the reports in the Everglades. He also pushed George Devlin Sr. to send cablegrams to everybody urging them to send letters and calls to their senators and congressmen, pressuring them to open a search in the Everglades.

Joan Powers had already gone to the Navy to pursue this. Believing implicitly in Navy incompetence, she unfurled a map of the reported flares in the Melbourne swamps and made her case about the ineffective search. Some Navy Brass eventually agreed to study it. She later admitted that they played "tick-tack-toe" over it trying to follow all the locations, and then, exasperated, asked her what she wanted. She was frank— she wanted the area searched more thoroughly and also wanted the name of the hunter who was out there. They retorted that the area had been searched, and the flares and the whole escapade was caused by nothing more than a "drunken frog hunter."

Fighting fire with fire had backfired for the Navy and Marine Corp. The press was still on the families' side. In a subject with its official conclusion still locked behind closed doors, there was reason to sensationalize. On the same page where the *Kansas City Star* printed the Marine Corps response, they also made no bones about the fact that "the mystery still puzzles the Navy." And with the legend mounting ever upward regarding a strange storm that sucked them to sea, the Navy was looking worse and worse by keeping their conclusion secret.

The press even buried some of its early buzz-kill theorizing.

At first papers speculated that the explosion "may have been the fate of the PBM search craft." But now the question was asked: "Did the Avengers, caught in the sudden tropical dusk, go down in a mass collision?"

This "storm" was becoming something quite absurd. The press said there was "a report by the aircraft weather station at Miami that a large area of turbulent air rolled out of a storm centered over Georgia, swept over Jacksonville about noon and reached Miami by nightfall. A meteorological freak— squalls on the surface, 40 mile winds at 1000 feet, and full hurricane of 75 miles an hour at 8000 feet— was recorded at 4 p.m. Such a development could easily have carried the torpedo planes miles out of their practice area." The press asked again: "Did a navigational error send the five planes off course, eventually to crash with empty gasoline tanks in the storm swept Atlantic?"

A legend was developing. "The first intimation of trouble came at 5:25 p.m.," read one column. "It was a radio message stating, 'have gas for 75 minutes more. Can't tell whether over Atlantic or Gulf of Mexico.' . . .The message also hinted the Avengers had been caught in a blinding storm . . .Subsequently, the air station heard pilots of the five planes talking among themselves. Most of the conversation was garbled, however. Then came the final message, 'I am lost.' "

The questions, the hypotheses, the speculation were mixed with the frustrating blackout by the Navy. Captain Burch continued to be bugged about progress in the investigation of these claims and in the swamp searches— two things the Navy was not even involved in since December. He responded to the press on March 10, 1946: "We have followed every clue clear through. We get new ones every day. Absolutely nothing has been found to indicate what may have happened to the planes." A reporter then told him that relatives were organizing to press the Navy to search. His response could not have been better even if he had been counseled in loopholes. "We will do anything they want us to do and will be glad if they can give us a clue to work on."

Burch, however, quickly torpedoed the whole idea of the men being in the Everglades. The press had quoted the Navy condemning it as "proved a cruelly false report." Burch put the final nail in the coffin. "There was nothing to the report." Then he rather interestingly qualified why. It was because Prob. Nav. 1 only required the planes to go out about 100 miles from Florida and was entirely over water. It was normally "only about a 2 hour flight to complete." He added it was "impossible" for the planes to have made it back over land in order to disappear in a swamp. "There are too many lights and too many air stations along the coast for them to have overflown it."

Due to the mounting controversy, Captain Burch even felt compelled to give a formal statement. In doing so, he completely minimized the weather. "What happened is unbelievable. Only 15 minutes before the squadron of Avengers left our base at 2:10 p.m. that Wednesday, another flight of five similar planes took off, flew exactly the same course and returned safely without incident. It encountered no unusual weather conditions, except the wind picked up 10 or 15 knots. In the previous two months, 250 navigational training flights had flown over the same area, without difficulty. And during the war, more than 8,000 such flights were completed, all without trouble."

The spotlight on Burch was a critical Navy mistake. It was putting the Naval base most confused over the incident in the front row of the Navy counter-attack. Unknowingly, by their rash desire to brush aside Gallivan, Powers and the whole controversy they were creating, the Navy had set their ship toward its first real iceberg— Katherine Taylor. Foolishly, they had made Fort Lauderdale NAS the weak bow of their ship.

Chapter 14

Gorilla Dust

JOHN GALLIVAN'S NEXT LETTER WAS TO KATHERINE TAYLOR. It was almost identical as the one he wrote to the Parparts. The difference was that Gallivan was now more confident than ever that the tracks seen in the Everglades were from the men. "Nobody has ever been in there and the guides should know what they are talking about. I would say it is a 50-50 bet that they are aviators. If the *Miami Herald* says the story is true, what more can we lose when we have lost everything now?" Katherine Taylor was ready to take the initiative. After all, Captain Burch had said he would act if given a good clue. She wrote back to Gallivan telling him she and her sister Mary Carroll were going to Fort Lauderdale to present the information in person.

The two sisters made the long bus ride. They crossed the selfsame Everglades in Southern Florida. Flashing past their bus windows was the dense wall of scraggly bayou trees. It must have been a dismal scene made much worse by the sweltering heat and the thought of her son starving miles beyond.

Captain Burch may have been dreading their visit. He had already received a letter from Mary Carroll reading how "dissatisfied" she was with the information that her family had received. Then she referenced in particular her nephew by marriage, Whitney Lowe. He was married to Charles Taylor's sister, Mary Georgia, and was himself a recently discharged lieutenant in the Navy.

Lowe had visited the base on December 13 to find out what had really happened. Although he had just recently been discharged, he went in uniform. He got better treatment as a result, but still got the impression he wasn't given the whole truth.

Some of the information he picked up, however, disturbed Katherine Taylor and Mary Carroll. To start with, he learned that Taylor had just received some telegram that upset him. Nobody knew what it had said, for he had put it in his pocket, so the story goes, and it went with him. Also, it was said that Taylor's own convertible Buick was found on December 6; it had been run into a ditch on the side of a road. Furthermore, Lowe found out that Taylor had said "I protest this flight" while leaving the Ready Room (or briefing room). He was supposedly joking when he said it, but what prompted such a remark and caused it to be couched in humor is not known.

So far, these were just loose strings. But Lowe eventually had a private meeting with Captain Burch himself. In this meeting Lowe was straight-out told to ignore the media version of what had happened. To Burch, the facts were simple. He explained everything to Lowe about the type of flight, the time when Taylor called and asked for help, and that foul weather was eventually coming. Burch's opinion was, naturally, based on what his staff of officers had told him about the fuel and ETA. As the incident was unfolding on December 5, he hadn't even been on base. His ignorance of everything is therefore pardonable since by the time of his meeting with Lowe on the 13th of December the search had not been fully evaluated yet; thus Burch could only repeat to Lowe the screwed up ideas of Kenyon. His recital was, obviously, *sans* any inference that the core officers didn't know what was going on.

In order for Lowe to get a firsthand picture of what he was talking about, Burch arranged for him to fly Prob. Nav. 1 with Lt. James Roy Jackson. Ironically, this was the same flight Poole had ordered after the Board grilled him on the homing beacon for which he had embarrassingly no answers. The Navy, in fact, was killing two birds with one stone but keeping Lowe ignorant as to the real reason for the flight.

The weather was brilliant, balmy and clear, Lowe recalled, and they flew with the canopies open. Jackson's story began when they passed over the Biminis. He pointed out the islands and said this is where Taylor got lost.

This rendition shows the influence of his superior, Charles Kenyon, since we know by testimony Poole did not buy into this. Jackson said Taylor then headed east following Cox's instructions to put the sun on his port wing and follow the islands until he reached the coastline— standard procedure *if* in the Keys. Jackson then said that Cox's well-meaning instructions unfortunately sent Taylor out to sea. Then to compound matters, Taylor headed northeast. That put him out over the Atlantic. That was that.

The extent to which the left hand did not know what the right hand was doing was evident, almost pathetically so, at this moment, for at this precise moment Jackson was assessing the strength of the base's homing beacon which disproved the entire Kenyon scenario. The Fort Lauderdale radio transcripts show that Taylor was at 4,500 feet when he reported broken land. If these were the Biminis, as Kenyon believed, he could easily have detected the homing beacon, as indeed Jackson did at only 2300 feet while over the same chain of islands.

Even without this damning bit of information the Jackson/Kenyon improv. staggers the mind with its stupidity. For instance, since the Biminis run north-south it would be completely impossible for Taylor to have followed them east-west. And if heading northeast, he would have to cross Grand Bahama, an incredibly distinctive landmass. And thanks to the radio fix Fort Lauderdale damn well knew the flight had gotten out of the Bahamas. In light of this, Kenyon's scenario is actually impossible since it cannot explain how the flight crossed over Grand Bahama without noticing it.

Now in March 1946 Katherine and Mary were certain that Lowe had only been "fed scraps" by the Navy officials. This may be the case, but it may have been a sincere meal nonetheless. Mary Carroll's letter left no doubt that they wanted more information and action.

There is ample reason to believe that by the time Katherine and Mary arrived at Burch's office, the Fort Lauderdale staff had uncovered the truth. This may have prompted Burch's

If the Navy's original scenario was put on a map its impossibility becomes undeniable.

distance when they arrived. Indeed, Mary Carroll reported that Burch's first words to them were "You shouldn't have come." His reason?— basically because Fort Lauderdale didn't have a copy of the report. For 2 hours Burch had "seemed most wary" and his "eyes and attitude expressed antagonism and a determination to fix the entire blame on one individual," observed Mary Carroll. This individual was "our son," Charles Taylor.

Then Commander Howard Roberts, the executive officer of Fort Lauderdale and a man who had been a Board member, was called in to clarify Mary Carroll's concern that Willard Stoll, the leader of Flight 18, had also encountered trouble. This question stemmed from what George Devlin Sr. had been told about two flights, with one abandoning Flight 19. Mary Carroll had even heard that the other flight had broken up and the students had landed at various airfields along the coast. Roberts could have been more categorical in his correction of the story, but he was rather cagey, superior and quite officious. "He completed the flight," he unctuously confirmed. Mary Carroll was to comment later that he entered "in a belligerent mood and remained antagonistic throughout the discussion that followed." Mary Carroll had the impression that at the "conclusion of the conference" they had "left behind two thoroughly annoyed gentlemen of the Navy."

This now brings us back to the original assumption that Fort Lauderdale's hierarchy discovered the truth. During the 3 hour-long meeting, Captain Burch had told the ladies something remarkable. He believed Taylor had misidentified the Bahama Cays with the Keys. To be tactful, but nevertheless allowing hints of his personal view that Taylor had cracked, he added that he did not know how an experienced pilot could have done that.

Burch's statement about the Cays is incredibly significant. Up until March, Burch and Fort Lauderdale had asserted Kenyon's scenario. Now Burch was finally coming a little closer to the actual location. By March it seems Don Poole had spoken up. If not Poole then someone else must have realized there was a gaff with Kenyon's scenario. And we must remember that his position gaff was inspired by the ETA gaff. Someone no doubt made a very unsettling discovery.

We already know that calculating Flight 19's position from the correct time of takeoff (2:08 p.m.) places them at the extreme of their flight triangle at 3:45 p.m., if on course. More significantly, however, this axes Bimini as the "chain of islands." The only other chain of islands that can now qualify is the Bahama Cays. Here they were outside homing range, IFF range, and radar range. In accepting this, we cannot accuse Taylor of refusing to follow procedure. He could only be justly accused of confusion, but Burch and perhaps Roberts realized Fort Lauderdale could be justly accused of incompetence.

Katherine Taylor and Mary Carroll were, in fact, the worst relatives to have come at the present time since they were the family of the flight leader whom Burch new damn well had been solely blamed by the Board. Now it would seem there could have been some ghastly culpability at Fort Lauderdale.

Commander Howard Roberts had no reason to be antagonistic otherwise. It is interesting to note that both Roberts and Poole had been in the same Torpedo Squadron 14 in the South Pacific. Both had also been transferred to Fort Lauderdale at the same time. It seems reasonable to assume Poole at one point confided in his superior that there was an error in the ETA. We know that Poole had calculated flight time endurance correctly by using the time of departure. Whatever had transpired in the last month or so, there is no denying the Navy at Fort Lauderdale was now deferring to Poole's estimates, which were quite correct. It is obvious that something had happened at Fort Lauderdale not long before the ladies arrived which made their visit even more unwelcome. Burch also had to contend with Edward Powers Sr. already, who visited the area for a couple of weeks and who also pointedly expressed that he was given the runaround. Powers had political connections, and if the ladies should find out some real controversial information, Burch knew they were in contact.

But if the officers there had no reason to be antagonistic yet with the ladies, they soon would. Katherine Taylor wanted action, and she knew as it stood now the Navy wasn't going to give it to her. She posted a $1,000 reward in the *Miami Herald*. Her words were intentionally inciting— an attempt to provoke the Navy to search again, an action which would turn out to be her own tragic blunder. She declared: "It just can't be possible that so many could disappear without a single clue. I shall never give up until I find out why. The Navy may have closed the chapter, but I have not. I still believe my son may be living in some remote spot."

Civilians took note of the reward, but no one ever headed out into the Everglades. The Navy took note as well. On March 20, 1946, while she and Mary Carroll were still in Fort Lauderdale, the Navy declared all Naval Personnel of Flight 19 dead. Official citation of death meant that a little more of an explanation had to be given.

The verdict had been in. Now it got out. The Navy explanation was a final one. It was one of blame— Charles Taylor was to blame. In their April 3, 1946, official statement, they added inexcusable error with truth. "At 6:p.m. he was directing his flight to fly east even though he was undoubtedly east of

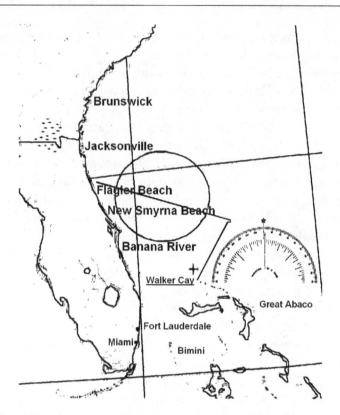

The most probable route and position of Flight 19. When Taylor said they were "over a small island" with no other land in sight they must have been over Walker Cay, the last isolated island of the Bahama Cays. They then went 030 degrees northeast, then briefly east, then west. At 6:15 PM they should have been off the Florida coast between New Smyrna and Flagler Beach. The strong southwest wind can account for their heavy drift northward. The circle represents the radio fix.

Florida. . .." They presented Taylor as confused to the point of derangement, giving orders that would take the flight into the Atlantic's abyss.

The Navy's verdict went out over the AP, and summaries were carried in many newspapers coast to coast. The families sighed mournfully. There was no question that there would ever be any search in the Everglades now. The Navy had done their job by accentuating the falsehood that the flight turned around and went east out into the ocean. There was no chance, in these circumstances, that they were still alive, certainly not in a swamp.

Even John Gallivan resigned to the fact that he had lost all he had in his son, Bobby. George Devlin, too, realized "Georgie" was gone forever; Smitty and "Rich" had accepted the little "snot nose" they had befriended at the age of 15 had survived the harrowing night patrols and the war to die off Florida. Mrs. Lightfoot had a terrible *déjà vu* of the news she had received when her eldest son, Eugene, went missing over France in 1944 and was then declared dead a year later. Nettie Stivers sat soulfully at mock services for her son while a soloist sang "Oh, Danny Boy." The Gerbers placed a headstone in memory of Jimmy in Minnesota.

Katherine Taylor was too sick to do anything. Mary Carroll was beside herself. This incident had been national and even international news. Now everybody had heard that one of the greatest aviation mysteries and tragedies of peacetime was placed squarely on Charles Carroll Taylor. The Navy had lost "our son," as Mary Carroll affectionately referred to her own nephew, because they were incompetent in their search. Mary Carroll wrote many vengeful letters to Naval officers involved. All of this was merely venting. It did no help and was agitating the Navy.

On April 10, 1946, Captain Benjamin Custer wrote an official letter from the Pentagon to clarify to the family members some of the "inaccurate information which various people have furnished you in the past" which he felt had caused a lot of the acrimony of late— the Everglades information and the original "drunken frog hunter" debacle. Mr. L.C. Smith's deposition was included inside, providing an at-face-value testimony that he had been the whole incident by waving angrily at Captain Morrison's plane.

The letter meant nothing to Katherine and Mary. While in southern Florida during their 9-day visit in March both had met with Morrison and had lunch. Morrison, though tactless in some respects, told them of the runaround he had been given and expressed his instinct that there had been a cover-up.

The Navy had clearly closed the case in April now, and was anxious to completely eliminate any other possibility than that the flight headed east under Taylor's deranged orders. Even when Senator Walsh, Chairman of the Naval Affairs Committee, made the formal request to search again the Everglades, the

Navy refused. The Secretary of the Navy, James Forrestal, was also petitioned by several congressmen and even by the American Legion, but all to no avail.

When Katherine Taylor recovered somewhat, she started the biggest battle through her congressman to get her son's name cleared. She fired off a very strong letter, to say the least, to Admiral Jennings at the Pentagon. In Katherine Taylor's gun was a very powerful bullet. Through the help of Congressman Milton West she actually had obtained a copy of the Board of Inquiry Proceedings in May of 1946. After a few months of agonizing reading and rereading, she could no longer stand it. Her son could not be the blame of it all.

Katherine Taylor was extremely cold and direct in her letter. She listed 12 counts, written up as though indictments, against Don Poole. They covered his reticence to send the Ready Plane while Cox was standing around asking to go. She could now cross-reference to the Board's proceedings, the witness testimony, the Opinions and Findings of Facts. The testimony of 3 other officers indicated they were being derelict, one was Commander Lawrence, commanding officer of Banana River, for being so slack about getting to base (he did not arrive until 9 p.m.), and even the testimony of Lt. Parmenter mentioned how hard it was to get Lawrence on the phone.

Admiral Jennings was surprised by the seriousness of the accusations. In view of these, he determined that the Office of the Navy and the Judge Advocate General of the Navy would review the report again. This was also taken more seriously when it was found out that Joan Powers and Edward Powers Sr., through Congressman W. Sterling Cole's office, submitted all the evidence they had accumulated in the hopes of starting a Congressional hearing (Joan Powers was the only other family member still holding out).

This was getting dicey. Jennings saw some of the Opinions of the Board to which Katherine Taylor was referring. One can well imagine that Jennings wouldn't want a public, Congressional hearing and probe with this kind of language being presented. The Navy would be foolish not to agree to review it in more formal (i.e. private) circumstances, which it soon did.

Amazingly, the Navy's review was shoddy, to say the least, and incompetent in the face of a possible Congressional hearing. This Review Board was made up of 4 Navy men ranging in rank from a mere Lieutenant upward to Commander, and the Marine was a Major.

This review started out with some insipid politeness, trying to clarify that an accusation of poor judgment was not derogatory to Taylor's name. From there it went to outright callousness. They had proven themselves even more imperceptive than the first Board when they said the Navy was not interested in assessing blame for the incident, but for determining the cause "in order to prevent similar occurrences."

The Review Board continued:

> However, the record clearly shows that the primary reason for the loss of the flight was the judgment exercised by him in allowing himself to be led to believe he was in a position in which he could not possibly have been. This is made more difficult to understand since Lieutenant Taylor had served in the Key West area for about a year. Through his confusion, broadcast to Lieutenant Cox, the planes were assumed to be over the Florida Keys until 6:10 p.m., at which time a position fix was obtained indicating that Taylor had been mistaken.

The Review Board obviously didn't review the original findings. No one believed Taylor was over the Keys. In fact, so certain was Poole that Taylor could not be near the Keys he didn't even bother to call Boca Chica and ask what their radar said. Had he done so he could have relayed to Taylor that he wasn't within 150 miles of the Keys. That would have convinced Taylor he was not in the Gulf. The Navy Review Board built up total guilt on Taylor based on the fact *he* said he thought he was over the Keys. Original sin was clearly being sought without any intent to address the fact that the original cause for the flight getting lost was not the reason why it never made it back.

This Review Board even took it beyond the previous Board, and decided to praise Poole. "In view of the type of ready plane available, the uncertainty about the lost flight's position, that a position fix was expected, and that weather was worsening, it is considered that Lieutenant Commander Poole showed good judgment and sound commonsense in refusing to send the ready plane."

Because Mrs. Taylor had made so many charges against the officers for incompetence, the Review Board became ugly and then smug in its reaction. "The charge of incompetence and neglect could better the laid to Lieutenant Taylor than to the officers who tried by every method available to extricate him from his precarious position. Lieutenant Taylor apparently did not cooperate, nor did he show good judgment in any of his actions after becoming lost."

In this Review Board's attempt to "objectively review" the case a second time, they actively went out and spoke to men and officers with whom the original Board did not bother to address. The most bilious information came from Taylor's roommate at Fort Lauderdale, Lt. Clark Miller. His nature to slander logic and the truth would become obvious too late (2 decades later), but for now the Review Board's adamancy seemed motivated by Miller's nonsense. He characterized Taylor as "haphazard and lazy" and his attitude toward Naval aviation as "somewhat indifferent." He said that Taylor had on a "previous flight" not even taken his earphones; also that he had made several requests to get out of previous flights "just to get out of them." For December 5, 1945, he was also seen boarding his plane without his plotting board. This Board could not conclude anything about that last point because the plotting board could have been in the plane. (Ironically, Clark Miller had been the nicest person to both Mary Carroll and Katherine Taylor when they visited Fort Lauderdale.)

Miller actually told the Review Board that he personally liked Taylor quite a bit. This should not have blind-sighted the Review Board to the impossibility of his claims. This was only Taylor's second flight at Fort Lauderdale. Miller had known him less than two weeks. How could he build up this historical dossier on Taylor? There was no way he could have come to his judgments in that amount of time from Taylor's one flight at that base, a flight on which Miller would not even have been present.

Miller's rubbish about a lazy Taylor caused them to dismiss Mrs. Taylor's point that her son did not want to fly because he could have been sick. They could "not give any serious consideration" to his excuse. "Since Taylor knew the proper procedure for being released from a flight if necessary, the fact he did not pursue the matter classifies his request as a whim." This, in fact, may be true. But pilot instructors often switched like this instead of going through the procedure. Willard Stoll even admitted that the leader of Flight 18 asked him if he would take his place, which he agreed to do. This actually later caused Stoll to think Lt. Arthur Curtis merely confused Flight 19 with Flight 18 and that Taylor never asked to be excused. If so, it is a dead issue. If not, the fact that Taylor took it as far as going to Curtis implies it was more than whim.

In conclusion, this board of junior officers got crudely psychological:

> Mrs. Taylor is laboring under an obsession which it is doubtful can be eliminated except by time or complete mental derangement. She appears to be incapable of accepting a rational, reasonable explanation of Lieutenant Taylor's responsibility for the safety of the flight. Unless she is told with finality that the Record has been thoroughly examined and is in order, she will continue to write letters and demand the time of officers who cannot possibly offer the satisfaction she seeks. It is believed that Mrs. Taylor is emotionally unstable as a result of this disaster and cannot be expected to react in the normal manner until she becomes emotionally stable.
>
> It is recommended that a firm letter be written to Mrs. Taylor advising her that the Record has been reviewed, interested parties have been consulted, and that nothing can be done which will improve the memory of her son. In fact, if the case were reopened there is every indication that Lieutenant Taylor's responsibility for the safety of the flight might be more firmly established and stated.

To compound matters Jennings, in his letter telling her the results, was crude enough to include that descriptive gem of her with the record of the Board of Review's conclusions.

Well, if you step forward they slap you on your arm; if you step back they bite you on the leg. The Navy didn't talk like this with Joan Powers because her father-in-law was influential in politics and she was going straight to Congress. It was time to get a lawyer.

To this end Katherine enlisted the help of an old buddy of

her son, none other than William L.P. Burke, C.C.'s old "Skipper" at Key West with Scouting Squadron 62. Burke was only too glad to assist in this endeavor. Burke never forgot that Taylor had saved his life twice while they were together in Key West. Burke was a trained lawyer and he knew the Navy. William L.P. Burke was perfect for the case. He had a personal interest in Taylor's reputation. . .and he wasn't going to charge them a penny.

Before Burke's aid would come to her son's defense, the House Naval Affairs Committee did in fact review again. Melvin J. Maas was the chief investigator. He was far more tactful. He conceded that there had been a lack of coordination, a break down in communications, and a delay in sending the Ready Plane. But he felt, based on the evidence, that even if these had been done correctly, it would not have brought the flight back.

However, Maas had labored under the same mistakes that the original Board had compiled from the Fort Lauderdale staff officers. He was unaware of the bungled ETA, the changing and inaccurate fuel estimates, the error in assessing Flight 19's position as around the Biminis, and with that assuming that Taylor had failed to use all the navigational and communication aids at his disposal. Maas had concluded just what had been concluded before: Taylor failed to act according to procedure.

Because the original Board was tentative even about Taylor's compasses malfunctioning, Maas was not going to go out on a limb. He thought Taylor's equipment *might* have malfunctioned, but glossed over the point. The original Board had based their skepticism merely on the testimony of the ground crew and the pilot who had flown FT-28 earlier that day. Since he had no problem with the compass, they rather illogically assumed Taylor should also not have had any.

Neither Maas nor any of the two Boards ever sought to explain in this context why Taylor was still crying for information on the flight's direction at 6:37 p.m. He was also plainly heard to ask Powers what course they were on several times. None of this would be necessary had he a working compass. The only logical deduction is that his compasses had ceased to function, just like he said. There was, incredibly, an unusual desire to blame Taylor, even to the point of denying or disbelieving his own words over the microphone.

The most likely explanation of why nobody really knew what was said is that the radio logs are placed pall mall within the body of the paperwork amassed, and they must each be read to get a semblance of accuracy and continuity of what was said. Maas, though well-intentioned, probably never read any of them. This certainly seems confirmed by his statement that he didn't feel that the shore units had any part in the disappearance.

This was the third bite in the leg. Although every bite had been from a different dog— the Navy Training Command at Miami, the Judge Advocate General's Office at the Pentagon under Admiral Jennings, and now Congress in the form of the House Naval Affairs Committee— the teeth had all been the same: the original "facts" as compiled by the original Board.

Mrs. Taylor would not give up. Her nephew Bernard Thomas had been an army pilot (a very decorated one) and was helping her to make heads and tails out of the original Board report. Thomas actually uncovered a number of discrepancies and conflicting testimony. In the margin of the report he even marked areas with "negligence"; "phony testimony"; "no transmission of fix to planes" (with 3 exclamation marks behind it); "hearsay evidence" and so forth.

It was clear to Katherine and Mary that a military man was needed, a pilot and lawyer. They decided to go to New York. Bernard's insights were real progress to them, and they had hopes that his 4000-word written analysis would re-energize Joan Powers and other families into action. There was still some political influence in Powers' connections, and this is what they needed since they felt they had expended all they had in Texas.

In New York they met first with Messrs. Devlin and Parpart separately since by this time neither wanted to speak to the other over some terrible disagreement they had had. Parpart thought Bernard Thomas' report meant nothing. He even disagreed with much of his conclusions. He was tired of the whole thing. Devlin was milder about it, but felt it was only henpecking at points. On the whole, it seemed irrelevant to him, too. Perhaps there was already bitterness because Taylor had been blamed. Clearing him would not bring back their children, and they saw no evidence that they were still alive.

The sisters then finally met Edward Powers Sr. in order to resuscitate Joan's "enthusiasm." Now they learned the truth about the last card they had to play. Joan had remarried and was even right now on her honeymoon. They knew it was useless to pursue the matter. Edward Powers Sr. told them of his own trip to Florida, his time and money spent to get basically nothing positive. He lobbied for investigation in Washington and then had finally given up.

New York had been a complete bust until they formally commissioned William L.P. Burke. Burke had warm memories of "Charlie." Burke was also a tower of legal enthusiasm. Lawyers say it once, then they say it a second time, then they tell you what they mean. He would handle the Navy. Burke told the ladies that a new office had been created, the Office of Corrections of Naval Records. This office was created for one purpose: there had been so many mistakes made during the war by the "90 day wonders" that there was a mass of paperwork built up by petitioning parties. (Whatever was the cause for really creating the office, the Navy knew there was a lot of room for correcting records.)

Burke was going to go before the Board in June 1947 and present the case to them. In the interim Burke played every angle like a good lawyer. At his artful cunning he even got nationally syndicated columnist Drew Pearson to write a very pro Taylor article describing Mrs. Taylor as a "soft-spoken, white-haired mother who gave up her quiet home life" to come to Washington and question the "brass hats, examine the records, to prove her son was innocent." Pearson even prompted public opinion a little by adding: "The new evidence indicates that inadequate weather reports and delayed rescue operations were responsible for the missing flight. She very much hopes it will clear her son's name."

Katherine and Mary went back to Corpus Christi. Burke had so impressed them with his sincere regard and happy memories of Taylor (and that he was doing this for free), that they felt they had a chance. The June meetings got postponed, then Burke postponed yet again. Finally, on October 14, 1947, almost two years after the incident, Burke entered room 2907 of the Department of the Navy. He opened his briefcase, placed his notes before him while the five-man panel waited.

Burke did everything but get on Fort Lauderdale's reputation with both feet and ride it around the room. He was all over the place. He mixed facts with error and, to be indelicate, outright BS. He divided between the cause of the flight getting lost, which he pointed out according to testimony was based on actual malfunctioning compasses; and the reason why it never made land again— the lack of coordination and outright incompetence at Fort Lauderdale.

Burke explained Taylor's excellent record as a pilot, how he saved his life, his record on the USS *Hancock* in battle, Rapp's praise of him in the Board proceedings which contradicted Miller's assessments by meager acquaintanceship, and other critical information such as the fact Taylor was told to remain on 4805 kilocycles. How can he be blamed for not following procedures about switching when he was told to remain on 4805 kilocycles? Taylor himself could easily have switched and picked up Air-Sea Rescue on the clear and strong channel of 3000 kilocycles, but what if one of the other pilots failed, or could not do so? After all, Cox's channel went out on that frequency. Then it would be dark and there was contact lost with another pilot, a student pilot! When it was too late and too dark he was told to switch. Taylor refused, thus gallantly jeopardizing his own life. Burke repeated that Taylor had ordered they all go down together, that they were to fly close formation. He was conscientious and tried to keep his flight together. Taylor no longer was painted as confused.

Continuing with other legal pedantry, Burke spoke a language which he knew the Board would understand. The original Board by placing the blame on Taylor had convicted him of manslaughter and/or negligent homicide in violation of Naval Law since Taylor had no attorney representing him. Burke destroyed the testimony of Lt. Cox who was, ironically, the only officer to perceive the serious nature of the incident early on and who also was the only man who had the right location and solution. He said Cox may have contributed to the loss by misunderstanding Cays for Keys and thereby sent Taylor the wrong information on how to get back.

Burke masterfully invoked the weather, to which end he could even use for his accomplice the original findings of Davison, in that it said Taylor did not take into account strong winds that blew him east of his course. There were therefore faulty Aerology reports, since such winds were not listed in the forecast for the evening.

In summing up, Burke said: "These transmissions supposedly from Flight 19 could be interpreted to indicate that the pilots did not have the slightest idea of where they were, in which direction lay safety, and, in short, were dubious as to which way was up. . . This is an attempt on our part to say the hop disappeared for causes or reasons unknown, which seems to be a natural assumption because nobody knows what happened to those airplanes. . ."

In his petition, Burke openly asked that the blame be placed on the staff at Fort Lauderdale, not on Air-Sea Rescue. The lack of coordination was not what he was really going after. He was going after the 90-day wonders and the "Naval Reservists" at Fort Lauderdale. The Board informed Burke that it was not allowed by regulation to comment upon the actions of others, but to review the incident and the findings. Burke therefore played the pedantic politics of the game and asked that the cause be placed on "conditions or factors undetermined."

Burke's dissertation was so influential it was clearly reflected in the wording of the conclusion that the Corrections Board would issue afterward. The Board forwarded its decision to the Secretary of the Navy on 28 October 1947.

> 1. The mother of subject officer, through her attorney, William L. P. Burke, submitted a written request that Lieutenant Charles C. Taylor, USNR, be exonerated and absolved of the blame for the loss of lives and aircraft from NAS, Fort Lauderdale, Florida, on 5 December 1945.

> 2. A review was held on 14 October 1947, at which time Attorney William L. P. Burke appeared and presented oral argument to the Board in subject officer's behalf. In addition, the Board also considered documentary material consisting of the record of proceedings of the Board of Investigations convened to inquire into the disappearance of 5 TBM aircraft which occurred on 5 December 1945 resulting in the missing status of Naval personnel and planes from Fort Lauderdale; subject officer's jacket, correspondence relative to the case, a brief submitted by Attor-

ney Burke, a brief submitted by Mr. A.B. Glann, legal adviser to the Board, and applicable statutes and policies in regard to the procedures and findings of the Board of Investigation.

3. After a careful and conscientious consideration of all of the factors of this case, the Board for Correction of Naval Records has concluded that the flight disappeared for reasons or causes unknown. Accordingly, the Board disapproves of that part of the Opinion of the Board of Investigation and of the reviewing authorities which places the blame on Lieutenant Taylor for the disappearance of the Navy planes and personnel. On the contrary, it appears from the evidence that some unexpected and unforeseen development of weather conditions must have intervened; that Lieutenant Taylor realized at an early hour that an emergency existed; that he promptly took charge, kept his flight together, and thereafter valiantly attempted to bring the flight home in the face of most difficult flying conditions.

4. The following action was taken by the Board in this case:

CONCLUSION:

(a) That an injustice is found in subject officer's record under applicable standards of Naval law, administration and practice, by reason of Lieutenant Taylor having been blamed erroneously for the unfortunate incident when there are no clear cut and convincing facts upon which to base such an opinion.

Acting Secretary of the Navy, John Kenney, approved their findings. A letter was drafted, based largely on the wording of the Board's decision, and was sent to each head of the families involved.

There it was. There was some truth, bull, interpretation, emotion and consideration of legal pedantry— "administration and practice" —in other words, the original Board did go outside of its authority and place blame, in essence, as Burke argued, convict Taylor without trial or defense attorney. The upshot was Burke had done it. The *Corpus Christi Caller* published Taylor's exoneration. By this time, it was but a filler. But it was done. It was all over.

None of the families doubted the results of the findings. The newspapers had been proclaiming a freak storm. "The first intimation of trouble came at 5:25 p.m." read one column. "It was a radio message stating, 'have gas for 75 minutes more. Can't tell whether over Atlantic or Gulf of Mexico.' . . .The message also hinted the Avengers had been caught in a blinding storm. . . Subsequently, the air station heard pilots of the five planes talking among themselves. Most of the conversation was garbled, however. Then came the final message, 'I am lost.' "

Therefore this conclusion was no surprise to the families. Nor was it any comfort. They knew their loved ones weren't coming back. Perhaps the only ones to believe it were Mary Carroll and Katherine Taylor. They still contemplated a trip to the Bahamas, in hopes of finding a lead on "our son."

The reader here would be surprised by how many members of the families to this day truly believe that some weird weather caused the flight's disappearance by blowing them out to sea.

With this excuse, everybody's tail had been covered. Flight 19 was the victim of some event beyond the control of the pilots. They were confused, everything was strange, odd. They didn't know which way to go. It is this scenario that would survive and captivate audiences in only a short time to come. Flight 19 had now truly disappeared behind the veil of the "unexplained."

Chapter 15

Flights of Fancy

WITH TAYLOR'S PUBLIC EXONERATION, WHAT ELSE COULD remain in the public's mind about this "unprecedented event in the annals of Naval aviation" but that some unforeseen "freak weather" had dragged the flight into the "Atlantic's limbo." The flight was then forgotten and the news articles, now yellowed, were thrown away. When the flight reemerged into the news only a few years later even Taylor had vanished. It was only the mystery that interested writers.

This article by E.V.W. Jones in the Associated Press was brief but captivating. Jones had some of the most peculiar mysteries of aviation at his disposal. No less than 3 airliners had vanished without trace since 1948, two near Bermuda and one while on approach to Miami airport. Adding these to Flight 19, plus a few ships into the mix, and the Atlantic's Limbo was easily narrowed down to an area off the Southeast coast of the United States.

In October 1952 George X. Sand would expand on the concept and call the area "a watery triangle" in his article published by *Fate* magazine, "Sea Mysteries at Our Back Door."

Individuals' names were not important in the recital of incidents, just the peculiar nature of the vessel or plane's loss and the fact that disappearance, not accident, predominated in this section of the Atlantic between Bermuda, southern Florida, and

the island of Puerto Rico. During the 1950s, locals even applied the colloquialism "The Deadly Triangle" to the area.

In 1962 a fascinating article was published in no less than the
American Legion magazine, written by Allan W. Eckert. It cap-
italized on the strange, freak weather conditions that dragged
the flight out into the Atlantic's deadly clutches. "The Mystery
of the Lost Patrol" was an exciting article because it offered
elaborate communication between the "flight leader" and the
base tower. Eckert claimed a high and reliable source in the
Coast Guard. The astounding dialogue was:

> Flight Leader: Calling Tower. This is an emergency. We seem
> to be off course. We cannot see land. . .Repeat. . .We cannot see
> land.
> Tower: What is your position?
> Flight Leader: We are not sure of our position. We cannot be
> sure just where we are. . . .We seem to be lost.
> Tower: Assume bearing due west.
> Flight Leader: We don't know which way is west. Everything
> is wrong. . . Strange. . .We can't be sure of any direction— even
> the ocean doesn't look as it should. . .

Well, there isn't much verbatim truth to it, but it was in
American Legion and thus it reached a large audience. These
sensational lines also gave impetus to many authors of otherwise
great credibility to speculate that there was some very weird
phenomenon involved in the disappearance of Flight 19 that
deserved discovery and serious study.

In and of itself this is a healthy approach, and it was in this
vein that most of the early accounts thereafter of Flight 19 were
written. The first to put it into a book was a former military
aircraft engineer, Dale M. Titler, in his 1962 compendium of
aeronautical mysteries *Wings of Mystery: true stories of aviation
history*. Quite a bit of chapter 14 "The Mystery of Flight 19" is
devoted to natural if not conventional theories of what could
account for the incident. But then "because every theory failed
to account for the weird sky conditions described by the pilots
and their failure to orient themselves. . .one possibility after another was tracked down and crossed off." He speculated instead

of "a magnetic disturbance from above" which has occurred in the past in "erratic, wavelike pulses."

Then in 1964 a journalist named Vincent Gaddis used Flight 19 and electromagnetic aberrations as the glue to stick a name on the area that would go down in history. His article in February's edition of *Argosy* magazine drew from earlier inspiration regarding "The Deadly Triangle," but this time Gaddis added to the moniker— "The Mystery of the Deadly Bermuda Triangle." The name stuck, and the whole concept captivated readers.

Gaddis followed up his article with his own compendium entitled *Distant Horizons* (1965), a popular book in which he devoted a chapter (The Triangle of Death) to the Bermuda Triangle. He reproduced Eckert's lines, adding afterward his own stunned reactions. "In all reports of this mystery, the importance of this last remark by the flight leader has been overlooked. The implications of an ocean that didn't appear 'as it should' are shocking. . . and chilling."

Gaddis continues:

Let's suppose that the patrol had run into a magnetic storm that caused deviations in their compasses. The sun was still above the western horizon. The Flyers could have ignored their compasses and flown west simply by visual observation of the sun. . . Apparently not only the sea looked strange, but the sun was invisible.

There is no question that such sources were respectable enough. But from there it blossomed. In full bloom the legend of Flight 19 had the pilots panicking, confused in the face of the unknown. All compasses were spinning; all dials were going crazy. The sun had vanished. The sky and ocean were turning strange colors of green. The legend of Flight 19 germinated the theories about electromagnetic aberrations and the Bermuda Triangle. Flight 19 was at the heart of it all. Eckert's dialogue alone was the thread that tied Flight 19 and thus possibly every other disappearance to a strange phenomenon of electromagnetic energy.

This led to a string of popular books on the Bermuda Triangle from 1969 through 1977 featuring Flight 19 (based on Eckert's lines) as the "first" and cornerstone disappearance. The most popular was Charles Berlitz's stunning bestseller in 1974, *The Bermuda Triangle: a saga of unexplained disappearances.* It engaged a worldwide audience and went on to sell over 5 million copies in hardback.

Because of its popularity, Berlitz's bestseller cemented this post *American Legion* legend of Flight 19 in the public mind. He even added to it with inferences of the paranormal, probing into the possibility that both Charles Taylor and another man scheduled for the flight, but subsequently excused, Marine corporal Allen Kosnar, actually had a premonition of disaster. Then, according to Berlitz, the flight roster was found to be complete indicating someone else "climbed aboard" at the last minute. Berlitz regards this "full compliment report" as a genuine mystery surrounding the flight, and in the Sun Schick documentary based on his book the narrator and host, Brad Crandall, even asks "Who. . .or *what* went with Flight 19?"

There is no need to accept the ephemeral mysteries in Berlitz's rendition. According to one of his later co-author's, William L. Moore, Berlitz had a knack for exaggerating minor points into large mysteries to heighten the popular appeal of a story. Kosnar himself has been off-and-on the fence about having had a premonition, most recently refuting he had any. He has said he simply didn't feel like flying that day. Today, he claims it was shrapnel he got in his nose during the war, and it was agitating him.[17]

But the "15th man" *is* a genuine sidelight mystery, and this nebulous entity has grown to cult status in the legend of Flight 19. Due to supposedly last minute roster shifts, quite a number of ex-military men are able to lay claim to the status. Eagle Bolotin is yet another contender. He loved flying and had flown that morning with another training squadron. He tried to get on Flight 19 as well. "But as luck would have it," he later said, "and luck is certainly the word, the same duty officer that had

---

17 It must be stated that in Richard Winer's 1974 groundbreaking documentary, *The Devil's Triangle*, Kosnar stated: "Two friends of mine, Bob Gruebel and Bob Gallivan, and I left the chow hall and went back to the barracks for just a matter of a moment's rest, stretched out on a bunk, and for some unknown reason I decided not to go on this flight."

been on in the morning was still on duty that afternoon and he recognized me." Since flying twice in one day was not allowed, Bolotin was axed. "If it had not been for that, I would have disappeared that day."

There is no need to challenge the veracity of either man. But neither could have been the 15th man. All gunners went on Flight 19; and Kosnar and Bolotin were both gunners. Actually, if we accept Kosnar's account of himself in Winer's 1974 documentary, *Devil's Triangle*, we know exactly who took his place. It would have caused Mrs. Lightfoot even greater anguish to know it was her son, Billy. By a letter to his mother, we know that Howell Thompson was scheduled (he mentions the flight for the next day), and Kosnar says that Gallivan was scheduled. Thus Billy Lightfoot stepped in for Kosnar.

The vacant position was that of a radioman. It is possible, however, since some aviators were cross qualified that Kosnar and Bolotin might have tried to go as a radioman. Given this, though, neither would have been scheduled; they could only have gotten on last minute as a replacement.

The simple truth, however, is that there is no 15th man. The radioman's position was being deleted from the Avenger anyway. If there was no one to act as radioman that day, there was really no need to replace him. When Powers was on routine maneuvers on October 9, two months before when he lost Belvito, he had no radioman. It was not crucial anymore. In all probability there was no radioman scheduled on December 5, and if the point was pushed, at the last minute it could have been any number of men, whoever was willing to go.

Perhaps the most bilious attempt to capitalize on the *concept* of the "15th man" was done by Calvin Shoemaker, who claimed there was a 6th plane in the squadron of Flight 19 and he was the pilot. Thus there was a 16th, 17th and even 18th man in the flight! Through an improbable rigmarole, Ensign Bossi and his crew went in their place.

Shoemaker was so certain his story was true that he contacted the Bossi family and conveyed part of it to them. Furthermore, when I inquired at the local historical society that serves Arkansas City, Kansas (The Cherokee Strip Land Rush Museum), Bossi's hometown, to obtain a picture of Bossi for my book and for an upcoming documentary on the History Channel, I was sent Shoemaker's account as a part of "historical" information about Bossi and Flight 19.

Quoting from his 3-page letter dated 3 July 1994 to "Mr. Marrs," I'll let Shoemaker recount his tale.

> On 5DEC45 six crews (three men in each airplane, except one which carried only one crewman)— a total of 17 men, briefed for the flight (Flight 19) and proceeded to man the airplanes on the parking ramp. Taxiing from the ramp were; leading Lt. Taylor, USN; wingman Capt. Powers, USMC; wingman Capt. Stivers, USMC; leading second section Ensign Shoemaker, USNR; wingman Lt. Gerber, USMCR; wingman Ensign Bossi, USNR (who was unable to get his engine started.)
>
> Lt Taylor and I both saw that Joe's engine would not start and noticed his crewchief removing the engine cowling. Joe and his two crewmen came running across the open field toward my airplane waving their hands over their heads signaling me to stop. Joe knew that I had already completed this mission on a previous flight and that it was not really necessary for me to perform it again.
>
> I braked my airplane to a stop and throttled the engine back to idle. Joe climbed up on my wing and shouted into my ear asking if he could take my plane and suggested that I could try to get his engine started and follow later to join up with the flight. I relayed Joe's request and the circumstances to lead Lt. Taylor. Lt. Taylor concurred and I, and my crewmen, got out of the airplane and Joe and his two crew entered and continued the taxi for take-off. This meant that Joe was now leading the second section with Gerber as his wingman.
>
> I subsequently could not get Joe's engine started, so I returned to the Ready Room and "Radio Operations" (just below the control tower) to hang out until the flight returned. . .

Being a classical student I will revert to Latin in an attempt to avoid offending the ears of the virtuous around me. Suffice it to say that Shoemaker's claims are pure *scattaurus*. Facts are easily accessible to contradict everything Shoemaker claims. There is a virtual paper trail on these aircraft. The most damning are the "Yellow

Sheets," the Daily Inspection Sheet for each one. They were saved and are a part of all the paperwork amassed by the Board of Inquiry. The pilot had to fill out these plane chits before checking out the plane. The chit was marked with the number of the plane in the upper right corner and lower left by the plane captain (head of ground crew). The pilot had to sign for the plane, date it, write the flight number down, the type of flight and then his crewmen's names.

Bossi's plane, T-3, the plane that he flew and vanished in, *was checked out by him*, signed in his own hand, timed at 1:45 p.m., the same time as all the other Yellow Sheets were filled out by the other pilots of Flight 19. Findings of Facts No. 7 "That Joseph Tipton Bossi, Ensign (A1)L, U.S. Naval Reserve, File No. ----- was the authorized pilot who signed for and accepted for flight TBM-1C, BuNo. 45714, for Flight 19 on the afternoon of December 5, 1945." Bureau Number 45714 is FT-3.

All planes were checked out in working order each morning. Those planes that did not pass inspection were not used for the day until repaired. Warm up for an Avenger is 10 or more minutes (usually). Big monster machines like Avengers simply didn't start and taxi away. Bossi's plane would have been turned over as soon as he signed the sheet. If it didn't start, he had more than enough time to get another. He would not need to run down the field and ask another pilot to get out spur of the moment.

As to switching, one physically could not even do that. Once you were assigned to a plane that was your plane. You simply can't play "musical planes." Also, more damning still, Avengers had no brakes. There was a holding brake. That is all. The pilot had to keep this depressed with his foot if he needed to hold the plane still; but if he let his foot off, the plane takes off. The engine would have to be shut off to allow pilots to switch.

Many officers testified before the Board of Inquiry. Anybody remotely related to the planes— crew chiefs, maintenance, training personnel— had to give account and explain the slightest variation in performance that day. How much more should someone who claims he was scheduled to take that flight, and even claims a discrepancy in starting a plane and a mad dash across the field saved his life? Yet in 500 pages of testimony and ancillary documents there is no Calvin Shoemaker.

The most unoriginal part of Shoemaker's tale— usurping Robert Cox's position— provides yet another interesting base for humorous fantasy. While waiting in the tower "We could hear Taylor better than he could hear us. In Radio Operations, there was an ADF 'birddog' instrument on which we received a weak directional signal toward Taylor's transmissions. I could hear Taylor talking with Powers. I heard Taylor— and Powers— calling in the blind for anyone to come up on the radio. I never heard Stivers or Gerber at any time on the radio." Later, "It was now after sunset, dark, raining at the flight's location and they were flying between cloud layers. The flight's radio signals were becoming weaker and weaker to Radio Operations— the birddog needle began to rotate with only a slight hesitation toward the received signal; then a steady smooth rotation."

> Getting no response from his last urgent demand for Taylor to turn around to a westerly heading, Joe radioed that he was turning to a heading of 270 degrees and "★----- if anyone wants to go with me, join up."
>
> Joe, Taylor and Powers ditched heading west before 8:00 p.m. Joe ditched last. I heard Joe declare his intentions before starting his flight descent on instruments several hundred miles on a magnetic bearing of 055 degrees from NAS Fort Lauderdale.

After this above fanciful concoction the letter ends with: "Please extend my best regards to John and the Bossi family."

The tale of Calvin Shoemaker proves somewhat unimaginative except for the entertainment value inherent in audacity. Unfortunately, however, it is not just an obscure letter contained in a small, rural historical society. He first told it on December 2, 1990, in the Los Angles *Daily Breeze*, entitled "Triangle's Truth." In this germinal story he also expresses what I believe is the motivation behind his novel claims. To be brief,

---

* Indicates profanity which Shoemaker redacted from this letter.

he seems to have such loathe for the idea of mystery and "magnetic forces" involving the loss of the flight that he has, wittingly or unwittingly, constructed a tale to prove them false. The article states how he "shutters" when they are mentioned.

But being the 6th pilot left standing on the runway isn't enough to debunk the sensational (or even magnetic) from Flight 19. He has to be in a place to hear the last dialogue and thereby confirm today that Flight 19 suffered a mere ditching at sea after getting lost, after mutiny and confusion. He even gives the exact magnetic bearings, which are completely apocryphal since Fort Lauderdale had no birddog. If they had had one they would have known where the flight was all the time.

The *Daily Breeze* article implies there's a lot of "unofficial conclusions" which bear him out, which do not appear in the report. The article states: "Shoemaker considers himself an authority on Flight 19."

Shoemaker was utterly obscure until this article allowed him to give himself the authority to dogmatically debunk The Bermuda Triangle. But even this brief sortie out into the world of Flight 19 wasn't much until five bogus Avengers were found in May 1991 ten miles off Fort Lauderdale by salver Graham Hawkes, and then unfortunately confused temporarily with Flight 19. Shoemaker was in Fort Lauderdale at the time and made a big noise at the hotel that he knew it was false. According to him, the clerk called ABC to let them know they had one of the pilots and an "authority" there. He got on ABC briefly, and was then filed away.

This landed him in some newspapers. The most tragic of which was the Arkansas City *Traveler*, which relied on the *Daily Breeze* article "that revealed more information about that day than has ever been told before." It then quotes, tragically, Bossi's mother Gertrude, the last parent of any Flight 19 pilot left alive. She had fallen under Shoemaker's spell. "He knew everything that went on that day," Gertrude told the *Traveler*. "I was so glad to get the information from him because I felt the truth of the story had never been told yet. So many bad things had been published, never the truth. That really touched me to find the truth of the story." Gertrude was 94 at the time and passed away shortly thereafter.[18]

Shoemaker is just one of many bit players who have surfaced as experts by having had some remote contact with the NAS or with a member of the flight. Bit players get published in local magazines, or have a walk-on on one of those documentaries of exiguous facts. One of the bit players, John Evans, claims to have been a gunner there and a part of the search. He had been seconded to base photographer. At the time Flight 19 disappeared he was in the dark room. Many still consider him to be in the dark room, which Evans proved in his article in the Florida *Gold Coast* Magazine in 1998. It was a careful sifting of spurious reports from old newspaper articles. In it he also says such utter nonsense as that Katherine Taylor periodically visited Fort Lauderdale until her death in 1973, with Taylor's picture in hand, asking charter boat captains if they had seen her son.

Like Shoemaker, Evans too presents photos to the press showing a group of guys standing in front of a plane. The claim then follows that members of Flight 19 are in there. Evans even claimed to have all 14 aviators in one photo and this photo appeared in a major magazine in 1991. Not one was in there that I could see. It is really quite laughable to see old guys pictured in magazines and newspapers playing with a plastic Avenger and showing photos of a dozen John Does. Shoemaker even obliged the world press in 1991 and pointed out the final position of Flight 19 on a map. He couldn't have come closer to the position of the USS *Solomons* had he been trying.

None of this, however, has stuck to Flight 19 like one enigma created so long ago. In 1972 this sensational mystery was in-keeping with the then-current hype over UFOs. Art Ford, touted as a reporter, lecturer and author, announced on a coast to coast TV show that in 1945 he had been told by a Ham radio operator that Taylor was overheard to say:

Don't come after me. . .they look like they are from outer space.

---

18 The information first came to the family and then to the historical society because of the *Daily Breeze* article. Bossi's sister, Bernice, lived in Santa Monica and cut out a clipping and sent it to the family. When I asked John Bossi, Joe's brother, whether he believed the story, I received a quick and definite yes. "He was there. I believe it to be 100 per cent accurate."

Ford then made much out of what his ethereal Ham radio operator had told him. In consequence was born the possibility that Flight 19 was abducted by flying saucers. This gained popular appeal, if only for media value. By 1977 the crewmembers of Flight 19 became the biggest part of the finale in Steven Spielberg's blockbuster film, *Close Encounters*, when they were released by the flying saucers. One of Charles Taylor's nieces, Barbara Lowe (Whitney and Georgia's second daughter), seeing it in the movie theatre without having had any forewarning was surprised when the character 'Fred Taylor' was obviously supposed to be portraying her "uncle Charlie."

Even if Ford's "source"— the Ham radio operator— did exist he could not have picked up Flight 19 over his radio. The bases along the northern Florida coast could not even do this with their apparatus. The frequency of 4805 kilocycles was used for very little, primarily the training frequency around Fort Lauderdale.

Trying to cash in on UFOs and the Bermuda Triangle phenomenon, Ford threw himself under the bus. UFOs were not even "discovered" until 1947 when, on June 24, Ken Arnold saw a group of unidentified craft near Mount Rainier in the Pacific Northwest of the United States. It is interesting to note that Arnold's first thought was that they must be some form of secret aircraft of the US Air Force. Yet Ford's account has Taylor decidedly aware of what something from outer space must look like two years before flying saucers even came into vogue.

Until 1975 the early popular accounts of Flight 19 appeared to be possible. For the sober they offered the possibility of electromagnetic aberrations. But like a chain, the legend of Flight 19 was only as strong as its weakest point. For Flight 19 this was not aliens and time warps but its own popularity. Sensationalism, when it becomes profitable, makes debunkery equally profitable. Sensationalism may capitalize on points of popular hearsay, but debunkery buries the truth after it destroys the embroidery. There is no objectivity in debunkery. It does not offer theories; it scorns and mocks what already has been offered. It breaks a weak link and therewith hopes it has destroyed the entire chain.

Sadly, Flight 19 fell victim to this, first in 1975 and then in

1980. Although the instigator, Larry Kusche, set out to find the truth, he ended up adding more error than the sensationalists. The only difference was he erred toward the banal rather than the fantastic. UFOs and time warps weren't the outcropping of this error; a sloppy, careless pilot and novice students were the new truth. Debunkers have a perverse lust for irony, and thus many "skeptics" ballyhooed his 1980 book *The Disappearance of Flight 19* as exposing the great cornerstone of The Bermuda Triangle as nothing more than the botch up of a bad navigator. Having the guise of an exposé, error toward banality can present itself as serious research, skeptical inquiry and sober deduction. And, indeed, *Disappearance* was hailed as a book written by a man who "set new standards for investigative reporting on popular topics."

In substance, however, this reputation actually hangs upon only one book. Kusche had achieved notoriety in 1975 as the man who solved The Bermuda Triangle largely because of a Harper & Row publicity stunt in titling his anemic book on the subject as having solved it. *The Bermuda Triangle Mystery— Solved* went on to sell well during the heat of popularity created by Berlitz's book and it placed Kusche on hundreds of TV and radio shows. In actuality, Kusche's book was scheduled to be out in April 1974, 5 months before Berlitz's book was published. Its format at this time was to be more of a compendium of 19th century newspaper articles. These were to be used as a sober counter-balance to contemporary writers' versions of the same historic incidents. This format would have allowed readers to balance modern renditions with the original articles to see how they have blossomed. While Berlitz's book was in accolades, Kusche's book went through an interesting metamorphosis. Kusche had said: "I departed from the original plan because I began to see signs that it was all more a lack of information than a real mystery that anyone had tried to solve."

In substance, however, Kusche's new format did not morph enough to reflect the new title *"Solved,"* but remained the result of only a 6 months stint of researching student questions while a librarian at Arizona State University at Tempe. Most of it was based on browsing old newspaper articles he obtained from interlibrary loans. It was, in fact, a typical sophomore level collection of newspaper articles and a few easily accessible accident reports, and none of the more official documents remotely solved the cases they covered.

Along with fellow librarian Deborah Blouin he had compiled *The Bermuda Triangle Bibliography* in about 1973, a collection of all the sources in the popular press he could find on Bermuda Triangle cases commonly mentioned. The existence of this work prompted authors such as Richard Winer, John Spencer and even Charles Berlitz to request copies from Tempe for their research for their own upcoming books. Berlitz even devoted special praise to Kusche and Blouin in the bibliography of his own book until, that is, Kusche's surprise book came out claiming to have solved it all and citing, in a number of cases, different source material than that given to the other authors. Berlitz was incensed. He then ordered the praise removed from later printings of his book.

This uncritical bibliography, however, was the surreptitious route by which Kusche had actually achieved publication as the expert of The Bermuda Triangle. When in 1973 Harper & Row also asked for a copy of *The Bermuda Triangle Bibliography* he told them he was writing a book. On the strength of only 2 sample chapters, they offered him a write-for-hire contract to complete the book in 1974. It is this dichotomy that was mother and father to the substance of *"Solved."* A selective number of cases were answered with often ambiguous epilogues by Kusche, outright wrong newspaper articles, and capped-off with a starkly dogmatic conclusion that the mystery was all nonsense.

Kusche's book, as egregious as its claim was, however, hurt nobody. But for *The Disappearance of Flight 19* his purpose seemed predestined to follow on the heels of his last success and "solve" it by finding something simple, if not elemental, to blame— in other words, exorcise anything unusual out of it, any breath of the legend; less charitably perhaps, find *someone* to blame. Paradoxically this required him to denounce how inaccurate the newspapers were unlike in *Solved* where they were the sacrosanct source.

The upshot of all this, unfortunately, is far from a trustworthy historical account. For instance, Kusche supplants the actual

Problem Navigation No. 1 and replaces it with a bogus one of his own concoction. "With the exception of Hen and Chicken Shoals, there were no positive landmarks over which the lead changes took place. Each student pilot knew only that when he took over the lead, he was to fly a certain heading for a predetermined time. If the navigation had gone wrong on a previous leg, he would not have known it, because when he was not flying the lead, he was kept busy flying formation."

This is shockingly inaccurate. The Testimony of Lt. James Roy Jackson, the Carrier Qualifications, Communications, and Navigational phase training officer:

Question 16 of the Board: "How do pilots on navigation flights determine or check the accuracy of their navigation while flying the problem assigned?"

Answer. "With one exception the termination of all legs touch a point of land and that one exception is within sight of a large island."

Relegating the real flight plan to obscurity, Kusche's amorphous flight plan provides him with a basis to show how five pilots could get lost, because without any landmark all it would take is one pilot to go wrong and the others would blindly follow. Kusche follows the above supposition and places himself in the cockpit that day. "The right hand controlled the stick, the left stayed on the throttle; continuous attention was required in the close group. Pilots flying in formation are generally so involved that not only do they not keep track of their position, they often do not even pay attention to the direction in which they are heading"— reassuring procedure for navigation training.

Question 18 of the Board to Lt. Jackson: "Does navigation instruction for student pilots point out the value of frequent comparison between mechanical and remote reading compasses?"

Answer. "In ground school when flying the link trainers and in the problems they work in ground school, it is brought out in each period the value of comparing their compasses at all times." Not only this, but each pilot was to know the headings being flown whether he was the leader of that leg or not. "The problem itself is worked out under the supervision of the instructor and each student's navigation is checked by other students and the instructor prior to departing on the flight." If the student leading a particular leg missed his landmark it meant one thing: all the pilots were not paying attention. The result is a failing grade for all. Kusche even quotes J.B. O'Brien (of Flight 22) talking about flying the exact course of Flight 19 a week later. "We didn't have any trouble except that one pilot flew the wrong way on a leg because the wind had shifted 180 degrees and confused him. I flew the next leg and brought us all back home all right, but the whole group got a down."

Contradictions such as this are ubiquitous throughout Kusche's thesis. Even his above statement about Prob. Nav. 1 having no positive landmarks is contradicted in other places by reciting the route, mentioning the landmarks, and then mentioning them again when he flew the route himself.

Lack of analysis also dominates his study of Charles Taylor, whom he made the centerpiece of his book. He is revealed through little snapshots from the memories of those who knew him during the war. He was laidback, carefree, very much a southern gentleman, meticulous about neatness, temperate; but when he got in the cockpit he was a "hell of a flier."

Statements from those who had known Taylor with Torpedo 7 include Judge Bill Brewer. He remembered him as "the coolest guy I ever saw. He couldn't be flustered. He was a damned good pilot and really knew the Avenger well." Robert Parish clarified he was "quite confident" and "wasn't a drinker like some of the others." Kusche quotes Joe Broadwater, Taylor's radioman: "When we came back from a raid, they'd always debrief us right away in the ready room. Then the flight surgeon would come in with those little airline-sized bottles of booze. Taylor'd always give his away." Norm Anthony was his roommate on the *Hancock*. His wife described Taylor as meticulous and clean and "always picking up after Norm." The praises go on and include how he got the Air Medal and how he was wounded and how fearless he was in ditching his Avenger three times.

On the other hand, statements from some officers at Fort Lauderdale paint him in the opposite light. In every case, however, it is from mere acquaintanceship. Willard Stoll was one

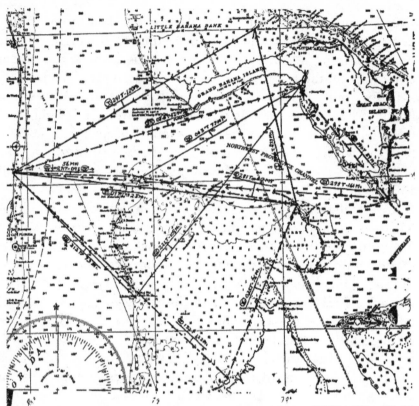

The actual Exhibit 1. The training triangles Prob Nav 1,2 and 3. Each corner is at a landmark, otherwise the purpose of the program was useless.

who claimed he knew Taylor well, but that he didn't personally care for him. He was not a bad pilot, he just didn't care for guys who drank all night and then slept late into the day. After 32 years (when Kusche spoke to him), Stoll appears to have been confusing Stivers and Taylor. Jerry Rapp even said Stivers and Taylor were similar but that Stivers was more outgoing. Stivers was a partier and heavy drinker, though he was a brave and highly honored veteran.

Kusche's analytical inabilities flowered when he came across Clark Miller's opinions. They were 16 years after-the-fact and second-hand at best, but Kusche gave credit to them. In 1961 Miller had spoken in depth to Lee M. Pearson, the Bureau of Naval Affairs historian. Miller, then a Commander in the Navy, apparently unloaded on Pearson. Some 14 years later, Pearson went over it with Kusche based on his firsthand notes. Some of what Kusche offers from these "notes" shows his ability to ferret red herrings. Kusche said Pearson characterized Miller as "quite bitter about the loss," especially the 1947 exoneration. He considered it a whitewashing of the man responsible. Miller characterized Taylor as "carousing all night and sleeping during the day," and even, as Kusche paraphrases, "given to taking naps on his flights."

Even a cursory examination of what Miller said would cause one to disregard it instantly. For one, Taylor and he had only been roommates for 2 weeks, and Taylor had only one flight in all that time (on December 2), hardly enough time and experience for Miller to form an authoritative impression. I have received the biggest laughs yet when asking TBM pilots "How do you take a nap in an Avenger?" After the laughter, the answer is predictable. "You don't." Even a pilot's inner leg muscles are used in flying an Avenger. If you fall asleep, you die rather fast in a deadly spin to the ground. That's the reason why Devlin used to sing in the intercom to keep "Smitty" awake on those horrendously long night patrols. If he snored off, even briefly, they'd had it. Miller creates Taylor as not just snoring off once, but having a proclivity to do so.

Kusche's inability to challenge patently impossible "remembrances" continuously places Taylor in a lazy or sloppy light. In quoting Taylor's comrades from the USS *Hancock* (in particular Bill Brewer) that were also transferred to NAS Miami, there comes the following story of how Taylor was as much *as 2 weeks late* in reporting to his new position at Miami, then when reproached by his superior just shrugged it off carelessly. "Bill Brewer remembers, 'Lieutenant Commander Spivey asked 'Where is Mr. Taylor? He's supposed to be here, too.' We had no idea where Mr. Taylor was; he'd left the *Hancock* months before we had. When Charlie finally came in two weeks later, [Commander] Spivey asked him where he'd been. 'We've had the bloodhounds out after you,' he said. . .Charlie sort of shrugged it off. . .' "

From his own personnel file, of which I have a copy from the Military Personnel Records Center in St. Louis, comes the truth: "April 11 detached Air Force, Atlantic Fleet [Norfolk,

VA], and to Naval Air Operational Training, Naval Air Station
Jacksonville, Florida, for temporary duty involving flying and
for further assignment. Reported 16 April. April 17 Ordered to
Naval Air Station, Miami, Florida, for permanent duty involving flying. Reported 18 April." Where's the two weeks?
Where's the tardiness? Where's the careless attitude? He arrived
the day after he was transferred.

None of these contradictory portraits of Taylor are questioned by Kusche. Strangely, there is no analysis. This becomes
even more tragic when he introduces alcoholism— not an accusation to be entertained lightly. He elaborates on the consequences if Taylor was hung over prior to boarding his plane
and the confusion that could result. More rumors of drunken-
ness are considered, and then nebulously negated when Kusche
defers to Willard Stoll's observation that Taylor was fine in the
Ready Room.

The fact that Kusche was *not* trying to label Taylor as a
drunk was lost on many. Nevertheless, his indiscriminate wan-
derings on rumors have caused Taylor and the Navy some of its
worst injuries. The image he left of a careless pilot is equally
groundless. One of his statements, based on general rumor, typ-
ifies the character of his Fort Lauderdale witnesses. "I also
heard that 'it was common knowledge at Fort Lauderdale' that
Taylor had been lost several times in the past. A few people had
even wondered whether he might not have been so proud that
after realizing he was wrong, he kept on going rather than
turning back and admitting his mistakes."

I mean, really!

Those "several times" he was lost are actually only twice in
his whole flying career. Once was in the Atlantic when he got
lost and then reoriented himself and made it back, just short of
the carrier. He had to ditch in its wake. The other was in the
Pacific when he failed to find Guam in a banged up Avenger
with malfunctioning equipment. Taylor himself had caught his
squadron commander in the Pacific leading them in the wrong
direction. What was it William L.P. Burke said? Taylor had
rescued him *twice* when he got lost out of the Keys. Getting
lost twice in 4 years of flying big oceans was not excessive. But,
again, it was only Taylor's mistakes that were put under the

spotlight in Kusche's *The Disappearance of Flight 19*.

One particularly offensive misuse of private papers given to Kusche by Taylor's sister and brother-in-law shows his weak scholarly approach. This surrounds the possibility that laziness and nothing else motivated Taylor trying to get out of his flight that day. Kusche notes that Burke's defense of Taylor ignored Clark Miller's claims that Taylor had merely gotten out of flights "for the sake of getting out of them." Kusche writes: "And Mrs. Taylor, understandably, did not recall an incident that had occurred several years before. On January 16, 1942, Taylor, then in training at Corpus Christi Naval Air Station, had told an officer in charge that he could not fly that day because of a stomachache. He reported to sickbay, then went to his mother's home to rest. 'Instead,' Mrs. Taylor wrote to her sister later, 'he took the car and insisted on going to the Chef to eat.' " Playing hooky *once* and Clark Miller's dubious claims are a weak brush with which to paint Taylor as a perpetual goldbricker.

Kusche's desire seems repeatedly to undermine the legend of Flight 19. Why else bring up all of the above (and more)? The result can only be to minimize the significance of Taylor supposedly asking to be excused from the flight. And this, of course, is the basis for the legend of Taylor's "premonition," another main cog in Flight 19's mythos. Yet in what appears to be his desire to further undermine the legend, Kusche introduces angles and claims that cancel each other out. For example, he underscores that Willard Stoll challenged the whole idea of Taylor trying to get excused. Stoll admitted he was actually not the designated leader of Flight 18. He took over for the assigned instructor when the instructor, feeling ill, asked him short notice. He thought that Curtis confused his flight with Taylor's. Which is correct? If Stoll is right, the legend of Taylor's premonition is removed, but so is the one piece of "evidence" that inspired others to "remember" Taylor as slovenly.

There is, actually, reason to defer to Willard Stoll. His Flight 18 took off at 1:45 p.m. Yet the legend of Flight 19 says that Flight 19 was 20 minutes late in takeoff. Presumably that's because Taylor was busy trying to get excused. If true, this would mean Flight 19 was scheduled for near the same takeoff time as

Flight 18. This is obviously impossible. Two flights cannot be flying the same mission and performing the same maneuvers over the Bimini wreck, an exercise that took about 20 minutes. They really should have been scheduled more than 20 minutes apart as well. Flight 18 was probably late for takeoff. Flight 19 probably took off on time, and Taylor probably never asked to be excused.

Illogical progression is also found in Kusche's use of "MT-28" to paint Taylor as disturbingly confused. "The fact that Taylor first gave his call sign as *MT*-28 (M for Miami), rather than *F*T-28, indicates that, at least at the beginning, he was thinking he had flown out of Miami, rather than Fort Lauderdale." Agenda is always made clear by someone's rank inability to contextualize. Contextually it is clear Taylor is using "Mike Tare" out of force-of-habit. He plainly says in the same breath that he is looking for Fort Lauderdale.

Many more illogical extrapolations are used to discredit the idea Taylor's compasses were ever haywire. "Taylor could have convinced himself that he was in the Keys if he had failed to pay proper attention to his navigation, seen some Bahamian islands he thought he recognized as the Keys, then assumed his compasses were wrong." This above deduction was a coattail to his dissertation that a "pilot's usual first reaction when he gets lost is to doubt his compass. It is tempting to have more confidence in terrain that looks familiar than in a compass that disagrees with the comforting landmarks."

In the case of Flight 19 Taylor could crosscheck his compasses with the four other pilots. The radio logs record a discussion on headings and compasses. Obviously theirs said something different or Taylor could not have concluded that his were wrong. Also, as late as 6:37 p.m. that night he is overheard to ask Powers "What course are we on now?" This incessant question between 6:15 p.m. and 6:37 p.m. (and perhaps later) is pointless if his compass is working.

Malfunctioning compasses are, perhaps, the greatest cog in the legend of Flight 19 and from there to The Bermuda Triangle, Kusche's prime target. He writes "All the men I have talked with who were in the tower say that nothing else was ever said about compasses. Taylor's brief statement, made immediately after he decided he was lost, are the foundations of the speculation that strange forces are affecting compasses in the 'Bermuda Triangle'." Actually, Kusche is shortchanging himself. Nobody knew Taylor's actual words until Kusche published extracts in *The Bermuda Triangle Mystery— Solved* in 1975. Nothing Taylor had actually spoken therefore could have influenced anything in the legend. Eckert's lines in 1962 started the speculation.[19]

An abrupt change of command is also a cog in the legend of Flight 19. This, too, Kusche questioned. He relied on the semantics of Lead and Command. "The issue very likely arose because of Opinion 34 in the report which states that 'Captain Powers assumed the lead of Flight 19 and maintained a course of 270 [west].' . . .It is not apparent, however, why this opinion was given at all, as there is nothing in the station logs or in any of the testimony from which it might have been taken. The situation is further muddled by Opinion 35, which states that at 6:06 Taylor *suggested* that Powers head east again."

Three radio logs clearly contain the word "suggest" and picked up the name "Powers." Opinion 35 did not make up the word or muddle anything. Commanders in the field do not suggest to a subordinate; they give orders. Had Kusche not been ignorant of Powers' seniority in rank, (his book dealt with the other pilots as mere bit players, if not outright walk-ons), he never could have made an issue of it. [20]

Kusche's rapture with fanciful memories and hearsay are clearly the origin for most of the erroneous ideas being peddled about Taylor and Flight 19 today. When I spoke with Don Poole at his home about the flight he brought out Kusche's book and swore by it, adding "I never knew about that stuff at the time." Based on Kusche's book, Poole believed Taylor had been a "bad apple" that ended up in his lap.

Remember, the officers all independently testified and were dismissed. The first conclusions they ever heard were from

---

19 Cox testified that he heard Taylor explaining his compass problems to others. Pilots Nathan Puffer, Melvin Pike and J.B. O'Brien also overheard his messages. Taylor also repeated his compass problems to Port Everglades when they first made contact.

20 I was apparently the first one to ever bother to get the actual records of each pilot. Powers clearly outranked Taylor by 5 months seniority.

Kusche in 1980, and these I feel were clearly mired by a pre-conceived goal. When Kusche found 3 surviving members of the original Board, including the smug Howard S. Roberts, he told them of Taylor's exoneration. This was the first they had heard about it. Kusche said they either laughed or were enraged— the obtuse leading the obtuse.

Georgia and Whitney Lowe, Taylor's sister and brother-in-law, freely gave Kusche, a man who called them friends, family papers, pictures and stories, only to receive in return a book that characterized C.C. Taylor according to second and third hand rumors, vague 35 year recollections, and then fingered him with the blame. Thus ended the friendship, and thus began a flight of fancy of Flight 19 that remains to this day. There is little to be gained from Kusche's farcical flight. An inaccurate characterization of Charles Taylor and the training flight, and even questions regarding the method of naval aviation, complete Kusche's false triad. "Everybody likes a good mystery," Kusche was quoted in a newspaper. "But to me, the really intelligent person likes the truth."

Something truly unusual did happen on December 5, 1945, which cannot be accounted for by vainglorious attempts to "solve" or steal thunder for personal reasons. No one can seriously study the disappearance of Flight 19 without carefully trying to figure out how they got lost, how Taylor could think himself over the Florida Keys when he was nowhere near, and, most of all, try and figure out where they could finally have ended up. It is time now that we consider much of what has happened, and put together recently uncovered points in order to assemble a clear picture of what we are up against. We must now begin a new stage in the research of Flight 19. Not one of putting in order its history, but finally one of trying to explain its fate.

42. Admiral Ralph E. Davison

Navy History Center

43. Admiral F. D. Wagner, center, with Admiral Callahan (left) and Fleet Admiral Nimitz. Coronado, 1953. *Navy History Center*.

44. The NAS's newspaper. Don Poole (left) and Howard S. Roberts (center) came from the same squadron. One wonders if Poole confided in him about Kenyon's mistake.

NASFLHA

45. Don Poole with his squadron, 1944, South Pacific.

Courtesy Don Poole

46. As Navy Liaison to Hollywood, Don Poole poses with Maureen O'Hara (for the making of *Wings of Eagles*). She was awarded "Miss Naval Aviation". Circa 1957.

47. Meeting Mitzi Gaynor for *South Pacific*, circa 1957/58.

48. Captain Don Poole at retirement, 1963. (all photos *Courtesy Don Poole*)

49. Corporal Allen Kosnar.
Was there really a 15th man?
NASFLHA

60. Whitney Lowe was one of the first
to question the Navy scenario.
Corpus Christi Historical Society

51. The "Jurassic Park" environment of the Okefenokee Swamp

All swamp photos from Author's Collection

52. Panorama of Chessar Prairie.

53.

54.

55.

Views of the swamp. One small section of the Boardwalk, the jungle view from an "open" section of it, and Grand Prairie.

56. An island in a field? In winter this would be flooded. Its appearance as a field does not mean it is stable. No closer view is possible.

57. Suwannee Canal, the main access canal into the swamp.

58.

59.

60.

58. Trees appear melancholy, hanging with Spanish Moss. 59. Gators are everywhere! 60. The dreaded ti-ti (tye-tye).

61. Bill & Rita Smith before the Memorial to Flight 19. Bill took the trouble to rise
from his wheelchair and stand to recognize his old buddy.
Author's Collection

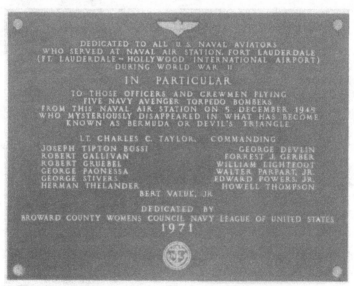

DEDICATED TO ALL U.S. NAVAL AVIATORS
WHO SERVED AT NAVAL AIR STATION, FORT LAUDERDALE
(FT. LAUDERDALE ~ HOLLYWOOD INTERNATIONAL AIRPORT)
DURING WORLD WAR II

IN PARTICULAR

TO THOSE OFFICERS AND CREWMEN FLYING
FIVE NAVY AVENGER TORPEDO BOMBERS
FROM THIS NAVAL AIR STATION ON 5 DECEMBER 1945
WHO MYSTERIOUSLY DISAPPEARED IN WHAT HAS BECOME
KNOWN AS BERMUDA OR DEVIL'S TRIANGLE

LT. CHARLES C. TAYLOR,   COMMANDING

JOSEPH TIPTON BOSSI	GEORGE DEVLIN
ROBERT GALLIVAN	FORREST J. GERBER
ROBERT GRUEBEL	WILLIAM LIGHTFOOT
GEORGE PAONESSA	WALTER PARPART, JR.
GEORGE STIVERS	EDWARD POWERS, JR.
HERMAN THELANDER	HOWELL THOMPSON

BERT VALUK, JR.

DEDICATED BY
BROWARD COUNTY WOMENS COUNCIL NAVY LEAGUE OF UNITED STATES
1971

62. The plaque on the memorial at Fort Lauderdale International
Airport, the former NAS Fort Lauderdale. Burt Baluk's name
is spelled incorrectly.   *Author's Collection*

Chapter 16

Clues to the Riddle

SPECIFIC INACCURACIES IN THE LEGEND ASIDE, THE MORE WE delve into the early moments of Flight 19's mystery the more we can appreciate the enigma that confronted its first biographers and the legend that they inadvertently fostered. Their accounts were innocent but sincere attempts to explain the chain of events as they were then understood. No writer knew that the flight had made it into the North Atlantic. Therefore all thought they must have disappeared in the Bahamas. The geography of this archipelago was as important as Eckert's lines in developing the legend of Flight 19. It underscored that *all* compasses on the flight must have been erratic. It was a deduction, not a sensationalistic lie. Titler's seminal account is vital.

If the patrol had flown west, it would have crossed the Florida coast or the Keys. Flying east, the pilots would have sighted the Bahamas with its twenty-five-mile-long Grand Bahama Isle. To the southeast were Andros and Great Abaco Islands, land masses impossible to ignore. Had the patrol struck out north or south over open water, the Florida mainland would have been visible at times. Despite high winds, the squadron could not have flown in one direction for an hour and 45 minutes— from the time they declared their state of confusion until their last radio call— without sighting one familiar landmark. The compasses of all five planes must have been erratic, for if the error was identical, all five would have flown a straight line and crossed land somewhere. Conclusion: the patrol flew in circles between the Bahamas and Florida.

Vincent Gaddis takes it further and shares his own feelings in *Invisible Horizons.* "Let's suppose that the patrol had run into a magnetic storm that caused deviations in their compasses. The sun was still above the western horizon. The Flyers could have ignored their compasses and flown west simply by visual observation of the sun. Apparently not only the sea looked strange, but the sun was invisible."

The actual facts (Flight 19 indeed getting out of the Bahamas) don't relegate Titler's and Gaddis' theorizing to sensationalism. The conundrum that inspired them is truly the greatest mystery of Flight 19. How did it get lost in this area to begin with, then get out, and not one pilot saw a familiar landmark or noted the position of the sun? These are very real enigmas that have to be answered and have been ignored until this book. They were addressed only in a very fluid way by the early writers, but they *did* try to address them. Because of the Bahamas' undeniable geography Eckert either came up with, or believed, those lines he first penned in his 1962 article. These lines may never have been spoken, but in general they do reflect a very real curio in the mystery of the "Lost Patrol."

Plain and simply put, to this day no one can satisfactorily explain how *four other* pilots did not mark the position of the sun or read their compasses in order to pick up on Taylor's navigational mistake and challenge him *immediately.*

Indeed the mystery of the Lost Patrol begins here. It begins long before the legend. It does not begin when Cox first overhears Taylor. It does not begin with Eckert's dramatic and fabricated lines. It begins in and applies to that period of routine silence after the flight turned northwest onto its second leg from Great Stirrup Cay (around 3:10 p.m.). It is at this point that Taylor thinks they were "going wrong" so he takes over and leads them back to what he thinks is the correct course.

Yet not one pilot challenges him. We know this because at
3:40 p.m. Taylor said: "Powers, what does your compass read?"
Then moments later: "We must have gotten lost after that last
turn." This is spoken very much in the vein of surprise. Therefore the subject of the course had never been raised by the others. Yet they must have been flying a wrong course for at least
30 minutes.

The silence of four other pilots is impossible to explain unless we agree with those who illogically assert that they must
have been on course and Taylor simply thought his compasses
were wrong. But this doesn't make sense at all and it is more
preposterous than the legend. At 3:40 p.m. they would have
Grand Bahama in sight if they were on course. Also, when
Taylor contacted Powers, four other pilots could still chirp up
and say 'nothing is wrong.' Yet Cox heard no such statements.[21] Rather he heard them talking about compasses and
headings, and around 3:45 p.m. Taylor even asked for suggestions.

There is some time discrepancy between Cox's watch and
Fort Lauderdale's clock. But be it 4:05 p.m. or 4:20 p.m., it is
about 20 minutes after Taylor asked for suggestions that he is
overheard again. This time he is reporting himself over broken
land. Based on the 5:50 p.m. radio fix, this broken land must be
the Bahama Cays, as all investigators have logically since deduced. Therefore in that 20 minutes Flight 19 got out of the
Bahamas northward without seeing land.

Something that stands out but doesn't make sense is an
enigma, but in a mystery an enigma is usually just a misunder-
stood clue. And this is indeed the first clue to unravel the mystery of the "Lost Patrol."

One other clue is needed. Ensign Nathan Puffer, up with
another flight, heard Taylor say he was heading north over
some islands. Yet the Cays point northwest. This would indicate a 30 degree error in Taylor's compass.

We can now begin to unravel it all.

To begin with, this 30 degree discrepancy explains how
Taylor got them lost. Taylor said he took them off course on

---

21 They also would have been over Grand Bahama at this moment.

the second leg. That would mean the northwest cross-leg of 346 degrees True. Because of the wind drift that day they were actually supposed to fly 317 degrees. With 35 knot crosswinds this would give them a course of 346° True. With a 30 degree incline in his compass, however, Taylor's would have read 346° when theirs read 317 degrees. He would have naturally believed that the pilots did not compensate for the wind. He would have taken them to 317 degrees (by his compass reading) which with a 30 degree incline in error would have actually been 288° True. Following this course they would have actually flown back *westerly* over their course between Bimini and Grand Bahama. Here both are separated by a 50 mile corridor of sea. With visibility at 10 miles they would not see Great Isaac's Rock and Chicken and Hen Shoals off to their left or West End off to their right.

The next important clue was actually uncovered by Kusche. Willard Stoll admitted to him that Taylor cut into his inner-squadron communications when he was chewing out a student. Taylor said: "Do your reprimanding back at base." Stoll was near Grand Bahama at the time, on his homeward bound leg, with his volume turned low. Stoll was sure Taylor had to be close by in order to have heard him. The same logic must apply to Robert F. Cox's first interception of communication. When he first heard Taylor at 3:40 p.m. it was merely inner-squadron dialog between Taylor and Powers. It was not being broadcast to others. Cox, too, must have been close to Flight 19. Within these two men's testimony, the solution to part of the mystery of Flight 19 lays. To be heard by both, Flight 19 had to be near both.

Cox had taken off at 3:15 p.m. and was still close to Lauderdale. When he first heard Taylor it must have been within 15 minutes after Willard Stoll had heard Taylor (and vice versa). At this point (3:40 p.m.) on a heading of 288 degrees Flight 19 must have gone so far west it was entering the Gulf Stream.

A basic chain of events is easy to guess. If on course the flight should be crossing Grand Bahama. But it hasn't. A landmark has not come into view. What would Taylor have done? First, of course, is to study his instruments and think of what

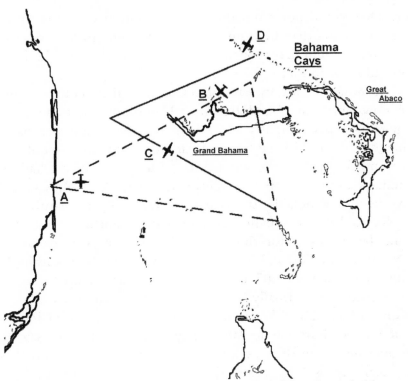

Dotted line represents Problem Nav. 1. Solid line is probable course of Flight 19 after Taylor took over. At 3:40 PM Cox was at position A; Stoll was at position B; and Flight 19 would have been at position C. Flight 19 ended up at D.

might have happened. Then a few minutes later he would call the lead pilot for that leg and ask him what his compass read. This we know happened at 3:40 p.m. "Powers, what does your compass read?"

The flight is continuing to head northwest as the pilots try and figure out what happened. Cox is now trying to raise them. They are moving at 140 mph. At this point they are soon over the Gulf Stream, slightly northwest of Grand Bahama. Cox clearly heard compass comparisons and headings discussions amongst the pilots. It seems almost unavoidable that one or more pilots would say: "I think we flew too far west"—which would have been the truth. They could indeed mark the position of the sun, but perhaps its position was not so drastically out of place that it would have gotten their attention earlier. They could point out the sun was more toward the tip of their port wing than high up; clearly that would mean westerly. That fact would have been undeniable. Taylor would have gone along with that.

The result is they fly east to get back to their course. They are now north of Grand Bahama. The southwest trade wind would have been slightly behind them and to their starboard. Their speed would have increased and their direction would be more northeast with the wind drift.[22] They would have had no landmark in view until a very broken "chain" of islands would have been unavoidable: the Bahama Cays. They are hardly like the Keys; they are too broken and sparse compared. But they could be mistaken for the broken island groups *west of* the Keys. With the tailwind, they would have been there within 20 minutes. It is now 4:05 p.m. Taylor has turned up his radio transmission and is finally in two-way contact with Cox.

One of the most overlooked words in all the radio logs is contained in Fort Lauderdale's. Taylor at first did not say he thought he was in the Keys. The first entry in Fort Lauderdale's log clearly reads "Nan How Able Three is in contact with T-28 (the leader of the 5 planes)— He believes he is south of Florida and west of Keys"— *west* of Keys. Taylor's erroneous conviction could not have been feasible unless the flight at one point had agreed they had headed toward the sun at some angle for quite sometime before they then headed east in response.

Fantastic?— yes. But it fits better with what Cox remembered. He remembered the inner-squadron discussions; the time it took for him to contact Fort Lauderdale and then try and contact the flight, again without success; then finally successfully contacting Taylor and hearing "I think I'm in the Keys but I don't know how far down and I don't know how to get to Fort Lauderdale."

The time periods used above follow Robert Cox's estimation. The Fort Lauderdale records indicate that Cox had communication with Taylor around 4:05 p.m., and that around 4:15 p.m. Taylor was reporting himself over broken land.

---

22 Stoll's flight must have passed behind them or abreast them in the other direction only minutes after (see map page 197). Flight 19 would have been largely silent heading east. Thus explaining why Stoll heard nothing from them.

There is always that 20 or more minute discrepancy in Fort Lauderdale's times as compared to Cox's. Splitting the difference just gives Flight 19 a little more time to have been flying westerly before turning easterly.[23] Stoll's testimony corroborates this somewhat. He had said that he was due back at 4 p.m. but was late due to some headwinds. While in the traffic pattern to land he heard Taylor report himself over islands. Thus it seems Flight 19 came upon the Cays by at least 4:10 or 4:15 p.m.

To summarize for a moment, we can deduce that the flight had thus flown in a westerly direction from around 3:10 p.m. (the time they would have been flying north from Great Stirrup Cay) to 3:50/55 p.m., about 45 or 50 minutes. At 3:40/45 Cox had heard Taylor call Powers. The pilots exchange ideas. They eventually head easterly. At least by 4:15 p.m. Taylor is over "land, but it's broken."

Since Taylor had no watch with him, he had to ask the others for the time. Accepting a certain amount of variation, he could assume that by 3:50 p.m. they had flown over their course perhaps close to an hour, enough to be 140 miles westerly before they turned and headed east. This is not enough to get west of the Keys, but Taylor could have thought the wind had changed at their altitude but not at a lower altitude; he could also have thought he caught a tailwind. Assuming that a change happened at his altitude and not at sea level would explain why the ocean swells had remained from the southwest. Such changes commonly happen.

The famous mystery of the Sahara, the B-24 *Lady Be Good* was the victim of this very quirk of fate when she encountered the still-unknown Jet Stream and overflew her destination in northern Africa until she ditched hundreds of miles beyond in the barren sand dunes of the desert.

Taylor's confusion is actually more astounding than that usually attributed to him, even by debunkers. He had actually believed they had flown further west than Key West and then in heading easterly came upon the island groups west of Key West. That's way over 200 miles west of their training triangle.

---

23 I also feel that Cox compressed time, as we all can do in recounting and event. Taylor probably called for suggestions after they compared compasses, not when over the Cays, as Cox's testimony implies. Of course, it could have been both times.

But to return to what we do definitely know about Flight 19 is to return to 4:05 or 4:15 p.m. when Taylor is now over islands and has his volume turned up. Cox and he are now able to hear each other and they exchange dialogue. Cox told him to put the chain of islands on his port wing and follow them to Miami. Good advice if in the Keys. But Taylor eventually came to nothing but that lone island and limitless ocean beyond.

"We're going too damn far north instead of east. If there was anything, we wouldn't see it."

Bay of Florida

Florida Keys

Dry Tortugas   Key West

Where Taylor thought he was and what he thought he was doing.

West of the Keys there are only the Dry Tortugas and the Marqueses. Following either takes one no where. . .except into deep blue ocean. This lone island must have reinforced his beliefs he had been west of the Keys.

In the Cays the "sun on the port wing" would bring one northwest to Walker Cay, which unquestionably must be the lone island Taylor referred to "with no other land in sight." The flight continues northwest past it until, now no longer seeing any land, Taylor orders them to fly northeast. Cox's instructions have not helped him. Taylor is sure he is in the Gulf of Mexico. As a result he thinks it best to head northeast, cut across the Gulf of Mexico and pick up the southern Florida peninsula near Fort Meyers.

This is not a drunken, careless pilot, but a pilot confused by his compasses and disappointed by the fact the other four pilots did not catch what he now realizes had to be a major navigational mistake after leaving Great Stirrup Cay.

The original Board actually called one of the pilots from Training Squadron 79M, Ensign John Sweezey, and asked him hypothetically what he would do given the scenario of Question 9. "If you had any reason to believe that your instructor was in serious doubt as to his position on an off-shore navigational flight, what course of action would you follow?" His answers are fairly typical of what his classmates Stivers, Powers, Gerber and Bossi would have done. In fact, he answered that he knew them very well. His answer merits quoting here.

> In the event that I, myself, had no idea as to our position, I would follow the instructor, but in the event that I felt sure in my own mind that I had the position, then I would question the instructor's navigation over the radio or by visual means, whichever was at our command, and try to work out something with him. We have always been taught to remain with the formation unless actual failure of our aircraft caused us to leave the formation.

Stivers, Bossi, Gerber, and definitely Powers, would also have challenged Taylor. Why didn't they have a clue? Why did they wait until Taylor notices something must be wrong?

The enigma remains.

Almost all those who have considered the mystery of Flight 19 begin by plotting its course at Walker Cay, the one position all have agreed upon. But the mystery of Flight 19 actually ends at Walker Cay. It begins here in the Bahamas during that silence after Taylor took them off course at Great Stirrup Cay.

After Walker Cay, Flight 19's courses are reported and they can be plotted. A new mystery begins at 6:06 p.m. Here precision stops. The only dialogue after this is Taylor trying to raise Powers and Powers possibly ignoring him. Taylor is heard to repeatedly ask "What course are we on?" After this Bossi and Taylor's call signs are the only ones heard. Whatever the reason, we can easily imagine Bossi is informing Taylor of what Powers will not. One interesting note is that Bossi was the only other Naval officer on the flight. The other 3 were Marines. (More than one Naval pilot at Fort Lauderdale has expressed the possibility there was a division between the Marine and

Navy contingent.)

Taylor's incessant question about the course in no way indicates Powers turned around. Why should he? They had been flying West close to an hour. They knew no land was behind them. Taylor would also not need to keep repeating the question. Taylor was either not being told, there was no course change going on, or there were many. Many seems unlikely.

We know at this point the flight was within the radio position fix. Using Walker Cay as a precise point and time confirms the position fix and vice versa. We know the forecast winds for all the altitudes up to 8,000 feet for the area were from the southwest, gusting to as high as 52 knots. We know the speeds of an Avenger. By 6:10 p.m. they should have made pretty much their full speed of 140 mph heading northwest (northwest because of wind drift). This would put them off New Smyrna Beach, perhaps 50 to 75 miles off the coast. This fits with the radio position fix report.

There really should be no mystery about Flight 19's continued course westward. The USS *Solomons* answers the question. Having flown this course, the flight would have actually flown within her radar range. There is no doubt about this. We have already discussed the carrier's position in a previous chapter, so there is no need to repeat that here. (Her sister ship *Mission Bay* and another baby flattop of the Bogue Class USS *Barnes* were also off Florida, but I have no positions for them.) Her one hundred mile radar range closes the door on the idea the flight turned around and headed east. She would have easily detected them.

But why didn't *Solomons* detect them coming into the coast? The simple fact is she did. If the flight had continued to head west, with the winds prevailing, they would have over-crossed the Florida coast near Flagler Beach. At this location they would have been at the extreme range of the *Solomons'* radar. They would have crossed this area around 7 p.m. coincidently just when the *Solomons* reported that unidentified flight of "4 to 6 aircraft, altitude 4,000 feet." At the extreme range of her scanning, the radar operator perhaps misjudged their direction. They would have been caught at the edge of the scope for only a short time. The *Solomons'* SC radar's hundred mile

The 5:50 PM radio position fix in relation to radar and radio ranges. The shaded area represents the only part within that position fix radius where Flight 19 would not be picked up by other controls.

range requires certain qualification. Ideally, this range would apply at 10,000 feet altitude for a modest sized plane. For a fighter, she had a significantly reduced range. However, the TBM was about in-between in size. Considering 5 closely grouped together, they could have been visible at the extreme distance of her range; and yet being at a lower altitude than 10,000 feet were indistinct as to how many.

It is a mystery why *Solomons* did not detect them sooner. The only explanation is that she was not monitoring her radar. But by 7 p.m. we know she was. And at that time we know Flight 19 was still up; and the only flight *Solomons* detected was near Flagler Beach.

Is this the case? Did Flight 19, in fact, overfly the Florida coast? The whole area was souped in with clouds and overcast.

At 4,000 feet Flight 19 could not have seen the coastline or anything below. If they proceeded inland, by 7:30 p.m. they might have been between Gainesville and Jacksonville, far inland by the time the dumbos were heading out to sea. Is this why nobody picked them up clearly at the end but Pensacola and Houma, both in the Gulf? Did Flight 19's antennas ice over the further north they went and therefore after 7:04 p.m. they could make no contact and could likewise not be overheard?

If that group of planes *Solomons* picked up was not the flight, what was it? Of all nights was it just a coincidence? There is really no room to argue. Even if one wants to winsomely believe that it was some other mystery flight, it does not change the fact that by 7 p.m. *Solomons'* radar was active. And by the known position of the *Solomons* we know that her radar was covering every place Flight 19 could presently have been *at sea*. Flight 19 *did not* go down at sea.

In order to underscore this more poignantly we must go back to that dark and confusing night 65 years ago and follow another unexplained report. Officers and yeomen were busily hurrying about at ASR. The time was 9:12 p.m. Teletype messages are coming in, the phone is ringing and radio messages are being handed to duty officers. One comes in.

ATC HAS JUST INFORMED BANANA RIVER THAT 5 UNIDENTIFIED AIRCRAFT WERE SEEN AT 8:50 P.M. AT 245 DEGREES 32 MILES FROM BRUNSWICK, GEORGIA, HEADING 150 DEGREES. ATC IS ALERTING ALL STATIONS

ASR quickly responded: "Can you ascertain if these could be the five missing planes off east coast of Florida?" An intriguing answer came quickly over.

WE HAVE BEEN GETTING THESE REPORTS FROM BRUNSWICK AND JACKSONVILLE. THEY ARE PLANES THAT ATC HAS NO RECORD OF. BUT IF THE FUEL SUPPLY IS CORRECT HOW CAN THE FIVE MISSING PLANES BE THEM?

Banana River's appended question to Miami shows the fatal

skepticism the ever-changing fuel estimate created, a skepticism that relegated further inquiry to a back burner. If any detailed follow-up was planned, it was quickly forgotten. Only a few minutes later both Miami and Banana River were consumed by a new mystery: the report from the s.s. *Gaines Mill* came in describing an explosion in the sky; the Navy was now busy with the mystery of what they would later discover to be the PBM Martin Mariner's destruction.

The upshot is that no more is heard of this unidentified flight. It never showed up at any base, though it was reported by two points, both Brunswick and Jacksonville. What happened to it?

Considering the precise coordinates given by Air Transport Command, there is only one explanation. This mysterious flight was right on the edge of the most deadly and impenetrable swamp in all North America— 700 square miles of nothing but an infested peat bog with billions of carnivorous bugs, thousands of hungry alligators, uncounted number of poisonous snakes; a place where the water is still but the land moves. Okefenokee it was and is called by the native Indians— in our tongue, "The Land of Trembling Earth."

To appreciate this possibility, and to come closer to a solution of Flight 19, we too must go to a place most people do not even know exists, to a place that appears to come out of the pages of prehistory, to a place that is jealously guarded by the Federal government.

Chapter 17

Don't Disturb the Alligators!

SINCE 1992 I HAVE BELIEVED THE FINAL RESTING PLACE OF Flight 19 was in the Okefenokee Swamp in southern Georgia. I felt it was more than coincidence that Air Transport Command should report 5 unidentified planes this night of all nights, December 5, 1945, in this area that should then also disappear.

In light of the true fuel capacity of a patrol Avenger, the ATC report can't be dismissed. The report also explains much. It explains why nothing was found at sea, but it also explains why Pensacola and Houma were the only stations that could best hear Flight 19 toward the end. Northern Florida and southern Georgia are definitely closer to these bases than they are to Fort Lauderdale.

Flight 19 alone fits those 5 unidentified planes. Flight 19 would be the one flight of which ATC would have had no record. Flight 19 would also be the only flight they too in turn could not contact to ascertain who they were (ATC also didn't guard 4805 kilocycles). In short, these facts mean those 5 unidentified planes weren't a scheduled, local flight, which would naturally be on or switch to local channels while in the area. What flight could this be if not Flight 19? As time went by and the search turned up nothing, ASR should actually have viewed this report as crucial despite Jenkins wandering fuel estimates.

Perhaps ASR felt it logical to dismiss the mysterious flight. After all, it never landed at a base nor was wreckage reported in any field nor did any aviator parachute down on any farm. This, however, does not negate the fact that this flight did exist.

Weather that night more than anything should have been taken into account before such a flight was dismissed. It was a moonless night. Clouds and rain plus wind had hit the area. The area is also sparsely populated and not well lit; most of the accessible area is wetland forests. There was really nothing for the pilots to see, and it should be considered likely that by this time of night they were low on fuel and were descending in preparation to ditch; one reason perhaps why the flight was spotted by two land stations.

Ditching, like a proper landing, you head into the wind. Smoke flares would have told the pilots the wind was from the southwest in the area. They would have to turn west/southwest to ditch. From the last position report this would have taken the flight over the heart of the natural abyss of the Okefenokee Swamp. Thus no other ATC hub would have been able to report the flight's advance. At that point it basically flew into oblivion.

Since 1927 these 660 square miles of peat bogs and impenetrable brush has been owned by the Federal Government as a natural reserve. The public is not allowed in. Flight 19 would have ditched in isolation in an area where nobody could possibly have stumbled across them. The ironic but logical upshot is that the Okefenokee is the one area where 5 planes could have vanished on that night, and it was the one area that was never searched.

With what I had, I felt it warranted an active albeit very tardy search in the swamp. My plans did not come to fruition for years—some 8 years after believing they could be in there, in fact. I was in for a lot of surprises. The first was the "hands-off" attitude of officials. When I called the Refuge and asked about wrecks, I made no mention of Flight 19. I only said I was looking for old military wrecks, dating to the time of World War II. Nevertheless, the answer was a dismissive yes followed even more quickly by a very discouraging "You can't go out there. If they are that old, they're

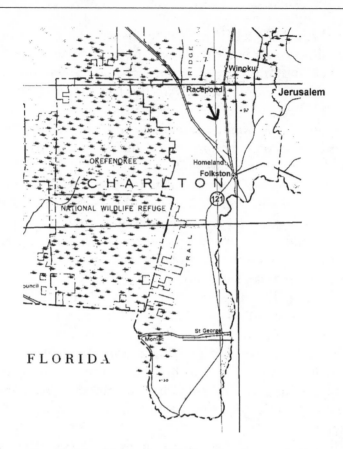

County map showing how huge the Okefenokee Swamp is. Arrow marks the spot where the 5 mysterious planes were reported by Brunswick and Jacksonville at 8:50 PM. The swamp is 38 miles north to south and 25 miles across.

all gone anyway. There's nothing left. The tannic acid in the water decomposes metal. After that long, there would be zip."

This was about the worst answer imaginable. As I continued in my preliminary investigation, it got little better. I was given the contact of Rhonda Kimbrough. She is the Program Heritage Director of all US Forests in Florida. She is also the only person to have excavated a vintage WWII plane (Curtis P-40) from a similar swamp. She clarified that "The engine would still be around, the propeller. . .There would be no human remains anymore, but there would definitely be something."

Rhonda sent me pictures of the Curtis P-40 excavation—the propeller was wrapped around a rock, the engine sat alone in a pool of water. From the looks of the Okefenokee, finding an entire plane was the same chance as finding a needle in a haystack.

It quickly became clear I had no chance of finding Flight 19 in person, but there was a very real possibility that the locals might recall coming across something. Despite the "hands-off" Federal attitude, shiners and poachers frequently got in there. Also, Federal refuge workers might have come across something a long time ago and never bothered to cross-reference any old wreck. Most of the employees are locals whose families used to own parts of the swamp, and most still live around the neighboring hamlets.

The immense publicity that Rhonda got for excavating that P-40 in 1990 (she was all over CNN) impressed upon me how carefully I had to proceed with even speculating that a plane from Flight 19 was concealed in a Federal Refuge. I was rather glad I never mentioned to any Fed what I specifically was looking for. So far, I was merely some wreck hunter whom they could dismiss.

However, a New York TV producer named Jonathan Grupper wanted to film me in the swamp and present some of my research on the History Channel. At this juncture I thought that might be the best thing, considering how stymied I was from Federal restrictions. I could at least present the case, and Jonathan was a very honest Emmy-nominated producer. Perhaps it might even open some doors.

This meant, of course, that I could no longer be stealthy in my expedition. Jonathan would finally have to broach the subject of my quest to the Refuge PR man, Jim Burkhart, in order to get permission to film. When time came, Jonathan told him flat out what I was looking for. Picture eyebrows hitting the ceiling while the rest of the face and body remains placid.

The reaction from every Fed was the same. "Nothing's left; you can't go out there anyway." The official stance aside, no Fed blinked at the idea that five planes could be swallowed up in the place. On the contrary, many felt safe enough with the Federal laws hanging over and protecting that natural abyss to express their sincere interest and even accentuate the possibilities.

A possible route of Flight 19 if they continued to head west. They would cross the coast near Flagler Beach. Then realizing they are over land, they might have swung wide to the north and come around. Compare this route to the December 5 search (shaded areas).

The news that there was a man coming out there with a TV producer for the History Channel following him looking for Flight 19 spread through the Federal employees of the area in a day and soon into Folkston, Georgia, the gateway town to the swamp.

Jack Mays is the former Mayor of Folkston and the official historian of Charlton County, a big chunk of which is the swamp. Jack was a score. He had owned part of the swamp himself when younger and finally sold to the Feds. I gave him a call. He found the whole story fascinating about Flight 19, and believed it was possible for 5 planes to have vanished in there in the circumstances I described. He was in the Navy at the time in Florida and recalled the incident vividly. He offered to track down a few names of people who could help me. He said he would play it close to the vest and let me know. (Jack would be a great partner in any subterfuge.) He finished with: "This is going to be *very* interesting."

Among a number of locals, Jack gave me the name of Cecil Carter. It seemed Cecil was the last of the true "swampers." This was good and bad. Good because Ceece knows the swamp; bad because it meant he was the last who had any real contact with the swamp's past. His help in locating a downed Cessna in 1978 had been crucial. He admitted to me there were other wrecks in there, but also thought that there could be nothing left from something as long ago as 1945. And the idea of anybody surviving in the swamp very long after a crash was unlikely. In the Cessna case, all that was found was the foot of the teenage girl passenger. The alligators did for the rest.

Despite the restrictions, my expedition had two very attainable goals: 1) Ascertain if it was possible that 5 planes could vanish in this area. 2) Find a trace of it in oral testimony, and possibly a physical trace that linked the planes to the area.

I flew to Jacksonville on the 3rd of August, 2000, days before the film crew would arrive. By the time I got situated, it was dark. A heavy rain was falling, but it was nonetheless hot and sticky. The weather conditions may have been identical to what was experienced by the flight on December 5, 1945, just a little warmer. The time was nearly identical as well to that odd report of planes— about 8 p.m. This was the best time to start the expedition. I had to see what the area looked like at this precise time.

I drove up to Folkston, Georgia. It's only about 45 minutes away from Jacksonville, probably a lot speedier approached than on a night so heavily soaked with warm rain and slapped by blustery wind. The small southern town seemed deserted even at this time of night. My headlights played off the weathered signs advertising the swamp and this led me to a road drawn in the blackest ink. South along Route 121 bushes and tree branches were dark roadside silhouettes lashing out at my Ford pickup like resident spirits angry for being disturbed. There were a few scattered houses and then nothing on either side: the blackness of wetland forests. Despite the stormy weather, I had to have air-conditioning on in the truck. The air is so heavy and sticky it was almost hard to breath.

Very soon I realized that what I had thought were wetland

forests to my right was actually the outer perimeter of the Okefenokee. The few, distantly placed streetlights didn't illuminate
much on a stormy country road. But silhouettes finally became
distinct enough for me to realize the dark mass was a curtain of
primeval-like Scotch pines. A couple of miles beyond them
possibly lay the object of my search.

On a rain swept night like this there would have been nothing for any pilot of Flight 19 to see from the air. In 1945, those
few homes that were in the vicinity still used gas lanterns for
light. There was no public lighting whatsoever. The Atlantic
coast is 40 miles away— that's 40 miles of mostly wetland forests. Much of it east of Folkston is owned by the International
Paper Company. This land is accessible physically (unlike much
of the mucky swamp), but at night it is just as abyssal from the
air. This is, in fact, one of the most unpopulated areas of Georgia, both then and now. South of this area is the Pinhook
Swamp in northern Florida. To the west are more wet forests
and sparse Georgian settlements. The few people who lived
around the Okefenokee back then would not have thought
twice about the drone of airplanes overhead. And deep within
the swamp there would have been nobody to hear anything no
matter what, especially in this weather.

After an hour of driving in the narrow, two-lane road, shining my lights here and there to see nothing but a thick wall of
spindly pines, I drove back to Folkston.

The dazzle of my headlights skimmed off the faded Okefenokee Swamp billboard again as I turned toward main street
Folkston. At the end of this street, just a couple of blocks
ahead, there was an imposing brick building with tall cement
columns. Its mean looking façade sneered at me. The stone
columns gave it an official air. I crossed railroad tracks and main
street and coasted along slowly. I turned to the right down another street and coasted past another old brick building. It
wasn't a tall building. It might have been a school at one time.
Now it was gutted and its windows were smashed.

This town is almost right under the exact flight path Air
Transport Command gave for those five planes. Being out here
on a night identical to that December 5 so long ago I could not
resist the temptation to stop on the street and get out and scan

the sky. The heavy humidity overwhelmed me and rain pelted my face. The sweet, sticky humidity creeps into your clothes immediately. You feel (and probably look) like a wet cat dragging itself around, now vapid and heavy with drenched fur.

It was silent, of course. The skyline of the town is still pre 1940s. At 9 p.m. on a dark and stormy night in 1945, the scene wouldn't have been any different at all from where I stood. The only thing missing for me were the running lights and the drone of the planes. Unusual moment.

It was time to head back to Jacksonville. I thought it appropriate to begin my search the next day, just like a real search should have been launched around here on December 6, 1945. I intended to be back up here early.

My search started at the Administration Building at a prearranged meeting with Jim Burkhart. Jim's job is public relations, and he does it well. A big guy with a broadcaster's voice, he is easy and eloquent. . .and a little worried Jonathan had prepared me.

The Administration Building stands beside the main entrance to the swamp near the end of a mile long paved road flanked with spindly Scotch pines. It is isolated and looks like an outback ranger station; it is new, spanking clean and official feeling inside— antiseptic, in a Federal sort of way. At least it was cool. Once inside the main door it was like hitting a curtain of fine, breathable air compared to the humid and tinted atmosphere outside. And I mean tinted, tinted with an unusual aura. The humidity acted like rose-colored shades for the sun. The area had a surreal, almost primeval lighting.

A lady sat protected beyond an admittance window. I felt like I was entering a health clinic. The walls have some nature material, pictures of the local natural inhabitants that must not be disturbed by anybody entering the swamp.

Jim came out from a closed hallway door to greet me in the large foyer. Then he led me back down the carpeted hallway. I felt like I was being admitted. He sat me down across from his desk and straight out asked: "Now, tell me just what are we looking for?"

I gave him the full nine yards, including the ATC report for that night. That an aircraft could disappear out there didn't seem to faze him. Every wreck out there vanishes quickly into the muck. But five aircraft raised his eyebrows. It seemed logical that one should have made it beyond the swamp. I explained Taylor's orders about all going down together. Jim then raised a good point about bailing out. I explained how Avenger pilots almost always ditched. And on a dark night they would not have bailed out over unfamiliar terrain. They may even have thought they were over the ocean.

We had a good conversation over some of the shoot's objectives. I felt at ease enough to ask him about his "big worry"—treasure hunters.

"Treasure hunters would be the worst case scenario," he admitted.

I got the impression that lesser scenarios would be media badgering and numerous public phone calls after the airing of the segment. Another one was that the Refuge has been portrayed inaccurately before.

"It is not a park!" he declared.

The Okefenokee is really not a swamp either; it is a peat bog. It developed over a sandy bottom after the oceans retreated at some ancient time. Cypresses eventually began to grow in the sandy, watery area. And as more peat built up, pines and every imaginable thicket and bramble started to grow. The off-scouring peat also formed enormous prairies covering dozens of square miles.

Much of the swamp, I was told, does look like land, but it isn't. That's only an illusion. Some of the "islands" are only areas where peat has built up into mounds. Pines sprouted up and took what root they could; but the islands, like the prairies, are unstable. You can stomp on some islands and the trees will sway.

These islands now look like pine-covered hillocks in the midst of grass-studded prairies. But the prairies are actually floating. They may be several feet thick, even 6 feet or more, but this relatively solid layer floats over perhaps 15 feet of water. Thus they can move and tremble under a man's footsteps. The Indians called it Okefenokee for this reason: "Land of Trembling Earth." Walk on a prairie and you can actually get vertigo. It's hard to tell how thick or thin some places are. If you fall through, it was too thin.

There are still a number of waterways, canals, and little tributaries that feed the area with highly acidic water. Because of all the peat, this "fresh" water is saturated with tannins and is the color of coca cola or a strong tea. It is fresh only in the sense it is not seawater. There is quicksand, there are snakes, carnivorous bugs, and where there are no "islands" or canals flanked by cypresses or floating prairies, there are dozens, perhaps hundreds, of squares miles of dense and impenetrable thickets called titi (pronounced tye-tye). That is the Okefenokee Swamp. It is not a land that time forgot. It is a land that time didn't want. Only the Feds wanted it.

I assured Jim I would do my best to make it clear that no piece of Flight 19 could be approached, much less retrieved, by any unauthorized expedition.

From his office Jim took me to what I call the "big board." It is on a tall wall in the very high vaulted-ceiling conference room. In order to reach every point of the map, Jim used a red eye light beam. He presented to me the absorbing history of the swamp; it quickly captures the imagination. The area is vast. There are some 121 miles of canoe trails through the swamp, yet the swamp is 98% inaccessible! These waterways constitute only 2 per cent!

In 1880 a lumber company dug some of these canals through the area in order to get in and harvest the virgin cypress (some of them 500 years old). Cypress is a slow rot wood and was highly sought after. Several companies tried to harvest the area but only went bankrupt instead. The labor involved in maintaining the canals and building railroad tracks to reach big cypress groves was not economical. On some occasions they would drive great 20-foot long poles into the ground as a base for the rails. In one instance, three 20-foot polls went in without hitting bottom. This continued until about 1927. Any trace of this era of foresting is gone except for some of the canals. Today, they are still used to access parts of the swamp.

With all the trees and vegetation, and the fact the ground of the area is actually flammable peat and not real "dirt," the Okefenokee has been subject to some horrific fires. In 1954-1955, for example, the swamp burned for 11 months; the fires ate up some 80 percent of it. In some places it burned huge holes into the peat, so that today little lakes can be seen from the air to call your attention to this past catastrophe.

Fire is the greatest worry the Refuge has. This is another excuse why people cannot be let too far in. Another is cost. It is very easy to get lost in there. One refuge worker got lost in a 1½ square mile area of it and it took a day to find him.

The area I thought most likely for Flight 19's ditching turned out to be the worst. It is square miles of essentially nothing but the thick swamp vegetation titi. Jim told me about a scientific expedition that tried to get in the general vicinity years ago, for various botanical and geological reasons. First, they succeeded in going about 12 miles into the swamp on the curvy boat trails. Then they tried to hike deep into the heart of the swamp. It took them 8 hours just to go a short distance. The prohibitive swamp vegetation and titi was so thick it was impossible to hack through. To continue would mean to crawl underneath, with the many species of poisonous snakes.

That's as far as that expedition got. And, as far as I know, that's the only one that even went remotely close to the spot I think most probable where the planes went down.

In fact, I don't think any real kind of expedition has been out into the swamp since then. Jim had been there for 28 years (at the time of our conversation) and admits that there are places in the swamp to which he has never been. "It's also a whole different place at night," he confesses. "Odd sounds of wildlife abound. We simply don't know all the species that are in there." Besides being vast, much of the swamp's growth, as described above, is simply impenetrable.

Because of the fear of fire, there is a Refuge helicopter constantly on standby. Twice a day it is scouting at treetop level looking for lightning strikes or smoldering fires. Since they also control the airspace up to 1,000 feet, which prevents the primitive, natural sullenness from being disturbed, a helicopter search would not be possible for me. That would disturb the alligators.

(The alligators are not disturbed it seems when there are only Refuge workers in the choppers).

Almost every afternoon and evening at this time of year thunderstorms come in, with the subsequent fear of lightning strikes. If I wished to pay 800 dollars an hour, I could rent the Refuge chopper. Something I certainly did not wish to pay! However, I would still not be allowed to fly under 1,000 feet. It would disturb the alligators. And if a fire was reported, I would find myself walking back to the heliport after being thrown out over the nearest road!

Jim took me over to a poster on the side wall. It was some kind of infrared thermal satellite photo of southern Georgia. The cities and towns registered as warmer colors of orange and red; the sparsely populated forests were deeper colors of greens to indigo; the area of the Okefenokee was marked by what could be called a black hole. Nothing was out there generating heat. It was pure wilderness.

Jim suggested I go to the "Boardwalk" to get a firsthand taste of what the swamp is like inside. I thanked him and left to drive there, after paying admittance.

Although it was a paved road, a lot of this drive is like a ride through Jurassic Park. Long spindly Scotch pines stick upward like long seedy bamboo from a bed of thick palm bushes. The air is warm, and the weak sunlight filtering through the heavy humid haze makes the palm bushes glow in an odd way like they were in dim artificial light in a photographer's studio. It was dreamlike and primordial. I could not believe there was a place like this in Georgia or anywhere in North America— and this is only on the very fringe of the swamp.

Even here, on the side of the paved road, there were small little natural ponds, sequestered lagoons, and flooded ditches. Invariably each one sported a pair of blinking eyes on the surface with a little mound behind it— each had a small alligator! Small long-legged birds "stork-walked" gingerly on the shallow banks.

The Boardwalk is a ¾ mile long, winding wood plank walkway cut through dense swamp vegetation, and it takes you straight into the swamp. It is a silent walk, except for an occasional birdcall or the stir of a crawling critter in the thickets somewhere. Visibility is poor on either side due to the dense growth. Some danger is truly evident. Huge spiders spin webs overhead and quietly watch your approach. Mosquitoes lurk in the dankest places, and with malicious intent swarm around you. I walked, swishing my hat around my head intermittently to keep them at bay. When they aren't around you, you can stop and listen: you will hear the low hum in the branches overhead or in the thickets over your shoulder; they are always near. You can't stop for long.

There is no water about. This walk shows what the entire swamp will one day be: an immense jungle. As the peat continues to build, the water will slowly be soaked up. The islands will be hillocks (as many are already). The prairies will eventually be real prairies and not floating ones.

Fortunately, some of this walk was under some of the heavy "jungle," so that it was cool for some of the way, though the mosquitoes attack you in the dankness of the shadows. When you emerge into a more open area, it becomes miserably hot, and even the mosquitoes leave you alone to dart back into the shade.

It wasn't hard to believe Jim's account of that scientific expedition that had to turn back shortly after leaving the open canals. It would be impossible to last long hacking away at the growth. No one would be able to walk the width of this place— 25 miles— and make it to the other side. Gators, quicksand and trembling prairies, carnivorous bugs, spiders and poisonous snakes aside; the titi and every other bramble would prevent one. Frankly, I could believe that the Feds really haven't got a clue as to all the species that live in the swamp; there may be a number of undiscovered or supposedly extinct ones in there. At night, like Jim said, the area is a bizarre orchestra of unusual sounds, animal fights, and eerie calls. You would not want to be on this path at night.

If it had not been for this boardwalk, there would be no way to ever see what is at the end of it. It stops at a beautiful open vista— Chessar Prairie. It is named after the Chessar family, who had lived in the swamp since perhaps the early 19th century, and they were basically the last family to leave in 1958. There is an isolated 50-foot tower that overlooks the prairie.

This is as close as one can get to a "bird's-eye' view of the swamp— and this is but a small section of the overall place. A cooling, soft breeze allows one to rest here.

I couldn't get over the repeating theme in the swamp— how new everything looked; how untouched by time. The Scotch pines were rickety and young, malnourished if you will. They might have been 100 or more feet tall, but seedy, thin. Then there were the fields of palm bushes underneath them. I swear I had seen such scenery in dinosaur books when I was a kid. I expected to see a Triceratops come charging through them or a couple of diplodoci thumping along. Where's the grazing brontosaurus? The only remnant remaining of their age was all the damn alligators sleeping about (fortunately none around here; no water). The scenery, the humid air, the rising warm vapors, the newness of everything, brought the image of a young earth clearly to mind. What did it look like 10 miles within this primordial peat bog? Again, it was hard to believe this was in America.

That night I picked up Jonathan at the Jacksonville airport. The next day we would meet with Cecil Carter and Jack Mays at the Okefenokee Restaurant, the hangout for all the "old timers." It was time to plot.

Jonathan and I stepped in and saw the man who must be Cecil sitting sequestered in a side booth by the dark corner of the place. It was right by the door, and everybody passing from that side isle went passed us. Cecil wore his tall baseball hat, tilted to one side. He looked like any American grandfather who had just been out fishing. He was genuine, relaxed and friendly. We sat down and Jonathan ordered us some lunch.

Ceece had been born in the swamp, one of the last locals to be so honored. He was proud of his "Okefenokee Accent," a very unique, subtle southern drawl. He had been a paratrooper in WWII and was glad he missed the fighting. After the war, he did some "hunting" out there, but as for going deep in he never did nor did he see any planes near the perimeter.

"Not even indications of a wreck?"

"No, you see if that happened in December the place would be really wet out there. No one went out there that time of year, not even poachers. By spring, the new growth would have removed all traces of any crash."

I soon got the full story about how Cecil helped the NTSB find the Cessna 172 in 1978. It had a young man and his girl-friend on board. After 5 days the NTSB still couldn't find it. From the last call it made while in trouble, Air Traffic Control knew the plane had to be in there; but so far there had been no trace of it. They finally had to go to tracking in Jacksonville and watched where its transponder code stopped. Then they flew over, and only then, with Cecil in the plane, did they spot part of its wingtip.

After Cecil led them out there, rescuers were able to dive down into the murky water and finally found a young woman's severed foot. The mother had only this to identify her daughter. That was all that was left. They never found the pilot.

Cecil had many anecdotes of being on the swamp when he was young, and was the first to express to us what would become increasingly obvious: the locals do not like the Feds owning it and frequently they let it be known. The locals had a true resentment of the Feds. The land was no good except to the Feds, so one by one the people sold what was their portion of the swamp. But they didn't like the fact it was being turned into a "forbidden city." Now they couldn't use any portion of it at all, and no one seemed to think the Feds were taking good care of the place. The whole place was like some deserted natural amusement park. All the locals are very proud of it, and they feel the area should get more publicity.

As we talked about the swamp, we had been targeted by a number of curious eyes. A hefty man's eyes had watched us the keenest. He re-donned his casual wide-brimmed wicker fedora and soon stood looming over us. He was Alton Carter (I don't know if he is a relation.) He is the true southern curmudgeon. He is a big gentleman in his 60s, big bear-like hands, thick forearms, dressed in denims and a nice blue polyester shirt. He knew who we were. The whole town knew I was there. Jack Mays had mentioned it on radio, and everybody knew Jack

Mays.

Alton scooted into the padded bench next to me. He leaned forward and crossed his strong forearms on the table. He patted it lightly with his attention-getting hand. His strong eyes shifted between mine and Jonathan's. He spoke in concerned undertones. "Just how are you gentlemen going to portray the swamp and us? That's what I want to know." Alton's first question really conveyed the local community's worry.

The Okefenokee, for all its size and uniqueness, is actually very little known. I had caused quite a stir, not just because a film crew was coming. . .but because there was a sensational reason why, and it affected the image of the swamp and that affected the image of Folkston. The famous Flight 19, which everybody had heard of, might actually be in there. This was disturbing as much as it was intriguing. The Feds were worried, and the town was concerned as well. I assured him this was no stunt, and that it was very possible the planes were in there. I understood that I would be prevented from making any physical search, but I needed to open some doors to that possibility. Alton seemed assured, though still skeptical.

Like Jim Burkhart, Alton thought one plane possible, but he raised his brow in a wry skeptical expression: "But five aircraft?"

I met his polite challenge in the undertones that our conversation had taken on. I presented the facts and the circumstantial evidence. At the end of our conversation, he was a bit less wary of the possibility. Alton was off on his way after thoughtfully inclining his head to one side. "Maybe."

The crowning moment had arrived when Jack Mays made his exuberant burst through the door. Cigarette in one hand and the pack in the other, he declared: "I'll be back in a minute. I just need to have a smoke. I wanted to let you know I was here first." Jack was back out the door.

Jack is the archetypical southern politician, a quintessential American invention. Jack is owner of WOKF in Folkston and another station in a town nearby. He's the town entrepreneur and swears the Okefenokee name is lucky. Among his enterprises is an onion company. He changed its name to Okefenokee Onions and by doing so wiped out his competition. He is exuberant, with a cherub's rosy face beneath perfect angelic soft white hair, a gleeful, toothy smile parting the rosy hue of his complexion. That's as close as any southern politician comes to anything angelic, and he loves to talk about the art of politics.

Jack was back in, and after saying hello to a few of the locals, he scooted in next to me and we four started talking. Now that the word was out, Jack didn't need to play it close to the vest. His announcement over the radio "caused quite a lot of interest," he confessed with a subtle expression. He couldn't find any shiners still living. It had been a long time since those days. He talked to Austin Hickox, the Mayor of Homeland. His father was a moonshiner in the swamp. He would be glad to come down and talk for the History Channel. He was the closest he could find. As for himself, he hadn't been in the swamp since he was a kid. The old Okefenokee was a bit of a mystery, even to Jack's generation. He had talked to the town's mechanic. He was a burly, roistering blade who had rescued people in the swamp. He was an alligator wrestler, and he would take us to his place of work next.

We got on to the swamp and on to the subject. Jack is not one for speaking in undertones. He brought up a crashed Navy flight in 1962. "Five men were on that plane. They couldn't find it for days. Once they did, they had to go down in there and look. Determination of death was made—" he started karate chopping his right wrist with his left hand "—by finding 5 right hands. There was nothing else left. That's how they figured they all had to be dead."

As the county's official historian, the first since "some lawyer up and died in the 1940s," Jack gave me some lowdown on the swamp. There was a botanist that went in there a few years ago to catalogue some of the flora, aside from that there's been no one that's gone in there for decades. An historian from the University of Georgia named Chris Trowell did historical research on the place; but as eyewitnesses to anything inside, there just weren't any. He did find the local pilot who found the Navy crash in 1962, David Glisson. He would come talk to me too, for the History Channel.

I was still at a loss to explain how both Brunswick and Jacksonville spotted those aircraft over Folkston area in 1945. Jack wasn't sure either. But during the war they had a civil air spotter's position up on top of the old courthouse during the war (that mean-looking building I saw my first stormy night there.) When Jack was 15, he was one of those assigned that station at night. "We thought the damn Germans were going to bomb us every night. It was a hell of a time after it was made co-educational. . ." I won't elaborate.

Jack was away in the Navy in late 45, so he wasn't sure if the post was still manned after the war. If so, that might have been one way the planes were spotted. But the fact both Jacksonville and Brunswick reported the flight smelled suspiciously to me of something a little more official, perhaps even radar.

After getting some ideas squared away, Jack was ready to take us to the mechanic's. He had a business appointment after that, so he would leave us there.

Jonathan and I followed Jack in his white thunderbird to the only gator wrestler around these parts. An open garage bay echoed with power tools. The area out front was jammed with precariously parked cars, as was the gravel curbside.

As I carefully tried to park the rental Ford pickup, Jack had gotten out and lit a cigarette. He smiled benevolently at my ignorance. "We don't give a damn pretty much how you park around here so long as you don't hit anybody." He ambled over to the noisy garage bay. We followed.

Out from under a jacked-up van came a smiling and friendly Pedro Drury. Jack introduced us while Pedro put down his power tools. Jack stood back, smiling and smoking. He would meet us later. He had some business to attend to and Jonathan and I were due to go talk to Bill Chessar next anyway at the local Baptist church. Church didn't look like it thrilled Jack.

Local Folkston mechanic and swamper, Pedro Drury, was another excellent contact, a representative of the younger generation and how the swamp continues to affect their life as well. He told us about wrestling with gators, but even he would not wrestle any gator over 6 foot. His point is made clear in person. He is a stocky, solid guy who when he gets tired of the phone ringing he just rips it off the wall. The "quickest phone changer around," he jests.

"Gators are cold blooded, so they don't have much energy.

They give it all they got in the first few minutes. If you can last that long, you'll make it. Some tourist got bit in the arm and must have passed out. You know, he was frightened or somethin'. He wasn't bad hurt. He woke up on the other side of a small lake. The gator must have drug him across. You see, they wait for you to rot and get tender and then they come back. . . but he woke up and walked out."

Deep in the swamp they are unavoidable, and some have been measured at 14 feet!

I asked Pedro what would happen to any survivor out there, *deep* out there. "They wouldn't stand a chance with all the gators, the snakes and just plain exposure."

The next evening I would bid farewell to the Okefenokee. I would hardly ever call my "expedition" a real expedition. I was not allowed to pursue any true search, and I was forewarned no aerial search would be allowed under 1,000 feet. That meant no infrared or thermal infrared could be used. I was, in fact, blocked from pursuing the possibility in any tangible way. I was a cautiously maintained special tourist. My trip was nevertheless the only sincere attempt to try and locate Flight 19 since the original search in 1945.

The Okefenokee is a giant. It can hide some of the greatest mysteries and it can materialize some of the wildest suspicions. No one remembered anything. There was no tangible clue. There was only suspicion on everybody's part that it was possible. One thing remained: that ATC report of 5 unidentified planes. I *had to be* on the trail of Flight 19.

This gave me no comfort. There was no triumph. Better that I was wrong. Better if it didn't go down in this acidic morass. Predators abound. The hunter is the hunted. Fire has scorched everything several times since 1945, burning the forests and the very ground. Fate, that finicky black-handed hunter that had made everything go wrong that dark night had also picked the worst place for anybody to crash in. Indeed it is better to go down at sea than in the Okefenokee. Politics aside, the Okefenokee was the ultimate enemy of my goal.

One must ally with another against a common enemy; and fire was the greatest enemy the Okefenokee had. Soil conservation photos must have been taken after the great fire of 1954–55, when 80 percent of the place was denuded of its growth. I began to wonder if aerial photographs taken after this time would not be my best chance to try and find a plane wreck. For any plane that crashed into the titi fields, this might be the only time they would be made visible; if it was in the mucky prairies it would take infrared. I had harbored a suspicion from the beginning that historic photos might be the only way to definitely establish if the planes were in there. I had to go back to old photos and come back to something new: infrared. I had to cheat Fate, the Feds, Nature, and even Time. I had to become an Infrared Detective.

Chapter 18

An Infrared Detective

LIKE A REAL DETECTIVE ON A CASE I SO FAR HAD A DATE FOR the incident. I had the "witness"— the ATC report. I had the circumstantial evidence— the fuel estimate mishap and the *Solomons'* radar return. My 'suspect'— Flight 19— fit the bill. I now needed the body.

The testimony I had to work with, however, was not from living witnesses. It was second-hand at best. Old aerial photos and infrared photos are like deaf mutes. I had to see and hear for them. I had to examine under a magnifying glass these impartial witnesses to see if they held any clue, even an enigmatic one like 5 funny blobs. The evidence, though it could be held in my hands, was paradoxically distant. It was like trying to solve a 60-year-old murder. A number of detectives and journalists have actually done so; but they were able to speak to witnesses. *I had to speak* for the witnesses.

First, I engaged Gary Anderson, an historical researcher into public archives in Washington DC. I told him my problem and told him that basic aerial photos may exist of the swamp showing wrecked aircraft. I believed this was a good bet from a conversation I had with local Okefenokee historian, Chris Trowell.

While searching the National Archives he had chanced upon

some pictures of the Okefenokee, and in one of them was something that ignited my hope. It was a photo of the eastern prairies. In it was a military plane wreck. Trowell had said the photo(s) looked impromptu, that is, it was not the mission of the aircraft from which the photos were taken to be hunting for any wrecks. It looked rather like the wreck was come across by accident, and the plane diverted so a photographer, who was probably aboard to take soil and topography photos, could snap a few shots, and then the plane continued on its mission. He thought it was an army plane, and that the photo was taken in 1943, too early for what I was looking for. But he admitted it had been 20 years since he had come across it in the archive's drawer.

That photo became more enigmatic as I searched the database of Army Air Corp accidents pre 1946. If it was a known wreck a good guess would mean it came from one of the nearby bases at Valdosta and Waycross. The records for all UAAF accidents are cross-referenced and indexed. Any crash near the swamp would be indexed in relationship to a nearby city or base, such as Waycross or Folkston, Georgia. Since Waycross had a military airfield during WWII, there were a large number of accidents indexed as belonging to Waycross AAFB or "x miles W of Waycross," etc. The eastern Okefenokee prairies (Mizzell, Grand, Chessar, Carter, Chase) would be southwest and closer to 20 miles from Waycross. This was the kind of annotation I began seeking.

Only one accident came close, a BT-13A Vultee Valiant, Serial No. 41-22590. This was a training aircraft that had crashed 12 miles southwest of Waycross. There were two other possibilities, and that was it. They were listed as Folkston, but were not given any other geographic cross-reference. One was a P-39Q, Serial No. 42-20797, a sleek pursuit Airacobra made by Bell; and the other was an A-25A Curtis (Shrike) Serial No. 42-79878. This was a big monster like the Avenger.

If it wasn't one of these 3, then I was on to a solid lead. Using a compass over a map, it was plain to see that the Waycross crash of the Vultee trainer can't be it. Only 12 miles from Waycross is still far too north and just at the edge of the forest area around the swamp. It was down to the two Folkston crashes. I had to wait now on specific dates for these. I contacted Mike Stowe, a veteran wreck hunter near Washington DC, who has a vast database of military accidents.

While I waited to see what he would turn up, I continued my pursuit of infrared. I had been referred to the USDA— the US Department of Agriculture in Salt Lake City— and I quickly sent a request to them for what they had covering Charlton and Ware counties, Georgia, the counties that contain most of the Okefenokee.

Meanwhile, Anderson's searches were providing me with disappointing information. He found that the photos contained in the National Archives were all taken from 20,000 feet! Even worse, these 478 stills covering the area dated back only to 1962. I knew it would be impossible to locate a plane by chance from that altitude so long after. They revealed only indistinct swampy topography. From that altitude trees looked like beds of mold spores. I would not even be able to see that Navy plane that David Glisson found by accident that very year.

It seemed logical to me that after the big fire of 1954-55 that soil conservation photos would have been taken. Yet another miserable irony was encountered— this turned out to be the worst time for aerial photography. Anderson discovered no photos could have been taken after the fire at any altitudes. Unfortunately, Anderson discovered from a number of cartographic staff members at the National Archives that aerial photography was in a sad decline due to the emergence of high altitude infrared. There would be no low altitude pictures unless like in the still-elusive earlier photo of the wreck in an eastern prairie they were taken only by chance during a routine flight that had some other purpose and also had a photographer— for the Okefenokee, that meant *never*.

I was not only disappointed but frustrated. I had publicly insisted that the swamp was the only place the flight could have been. Nothing had changed my mind. I still had the evidence, but I was beginning to worry about finding incontrovertible proof. Friends like Bill "Smitty" Smith had also told TV crews and interviewers that he believed I was right that Flight 19 ended up in the Okefenokee Swamp. I couldn't let down others.

I was left with infrared, which frankly I thought from the beginning would be my best chance. Infrareds, however, were not turning up. Anderson learned from the National Archives aerial expert, Gary Luchansky, that no collections exist at the National Archives consisting only of infrared photos. Also, even more devastating he knew of no archive that had any for that area. Infrared maps did exist, but they would be indexed along with overlay photos.

My leads with the USDA in Salt Lake City at first seemed more promising. They confirmed they had infrareds and sent me a Xerox of the flight paths of their photographic missions. Then came more disappointment. Incredibly, the earliest photos they had were from 1983! I unfolded the map to discover that the flight paths of the photographing airplane followed strict parallel lines about 10 miles apart. This meant that most of the area, the very heart of the Okefenokee, was in the gap between the flight paths. It had not been photographed; at best only the northern part of the swamp got photographed and most of Pinhook south of the Okefenokee. Even worse, the photos were taken from 58,000 feet overhead! It would be impossible to detect anything.

Thermal infrareds would be the best way to detect anything in the swamp by being able to detect the reflection of warm Georgian sun off an old wreck in the undergrowth. However, now I found out in order for them to be any good they had to be taken from a flight around 250 feet up. Over the Okefenokee? Never. My depressing hunch was confirmed by Gary Anderson. The lowest altitude photos of any kind were taken at 20,000 feet. These dated to 1987 and like the others above only covered parts of the swamp.

It was obvious Flight 19 could not be found this way. Amazingly, hundreds of square miles of Georgia have *never* been photographed, whether from high or low altitudes. And that which has been photographed showed bland rumpled and distant mottled topography— forests and swamp.

The Okefenokee's legal power extended 1,000 feet, but its infamous reputation as a morass extended upwards to 20,000 feet. Even official photographic missions didn't think there was any reason to fly lower over the area. The mute witness of aerial photos turned out to be blind as well.

My months of infrared detecting weren't taking me up a completely blind alley, however. Far from disproving my theory, the absence of any hint of civilized interest in the swamp seemed more or less to explain how this could be the only place where five planes could vanish. There are, oddly enough, clearer pictures of Mars' topography than there are of the Okefenokee Swamp.

I began to think that that enigmatic photo up at the National Archives would be the last ironic chance of finding a concrete lead on Flight 19 in the swamp. This photo became more and more tantalizing with Mike Stowe's response. He was finally able to confirm the location of those aircraft wrecks in Georgia. Both of the Folkston crashes were at Folkston. The first was on August 27, 1943. The pilot spun in and crashed. The other was on November 19, 1943. It nosed over the field and crashed, with only minor injuries.

Thus it was now clear to me there was no Army Air Corp wreck in the eastern prairies. Just where did that plane come from in the photo?

Fate, that fickle, black-handed predator that has a perverse lust for irony continued to dog Flight 19. I had sent Gary Anderson to look for it. Weeks later he confirmed it would be extremely expensive to try and locate it since Trowell could not even remember a group number or even the precise branch of the archive. The chances were slim of finding it for less than thousands of dollars of research time.

There I was content to leave it, with the implication hanging out like the ugly tongue of a gossipy Dickensian hag. But in April 2005 Larry Landsman contacted me through Patrick Huyhge, one of the editors at Paraview. Larry is the special project director at the SCI-FI Channel in New York. They had a miniseries coming up in December entitled *The Triangle* by Bryan Singer and Dean Devlin, and he wanted to capitalize on it with a special documentary based on the man (me) who really did investigate The Bermuda Triangle. To make a long story short, I told Larry of my theory and how stymied I was with the swamp restrictions and infrared. He requested this MS and used it to pitch a huge documentary to be produced by NBC News Productions.

Everything seemed to be coming my way. NBC was going to shell-out to swing into place the highest declassified satellite and take a 5 spectrum picture (thermal being only one) of parts of the swamp I marked as most probable (even NBC couldn't afford to photograph the whole swamp). Although by this time I was skeptical about the power of infrared to turn up anything, Jon Schriber, the senior producer, encouraged me with the fact that this satellite lens was capable of zooming in on a two and one half foot wide area of the swamp. Based on the last report of those mysterious 5 planes and wind drift, I marked the central southern area and then the eastern prairies.

In addition to the satellite search, NBC planned on looking for the Martin Mariner. I had pitched this idea to them as well. I viewed the Mariner as a "sitting duck." It was in only 78 feet of water and I had the exact coordinates from the *Gaines Mill*. If the swamp failed, the finding of the Mariner would constitute a major news item and would be a great promo for the documentary. Schriber obtained David Bright to command the naval expedition. Bright had just finished diving on the *Titanic* and was in decompression.

Clear skies finally came in August, so the swamp could be photographed. Weeks went by, and then the verdict came in as clear as the water of the Okefenokee: "nothing suspicious." It turns out that many areas of orange blotches were detected, indicating metal below. But nothing apparently to indicate aircraft was identified. I reminded Keith McKay (the producer who was in charge of the swamp aspect) that the swamp was once a place with many railroad tracks, stills, and other known wrecks. Still, he couldn't get anything else out of his contacts beyond the initial response: "We have not seen anything suspicious in the imagery." The gloominess of this statement is mitigated only by how unqualified it is. I know that the areas where known wrecks exist (e.g. "the 1978 Cessna" or the mysterious wreck) were photographed and yet apparently show "nothing suspicious."

"Nothing suspicious" really means two things— not even the mysterious wreck in the eastern prairie or the "1978 Cessna" left imaging that registered as an airplane. . .or it meant that those examining the readouts didn't know what to consider the image of an airplane wreck after 60 years under the corroding peat and water of such a swamp. Without consulting known wrecks, how could one compare?

Worse news than this was next. For 7 days Bright had led two vessels, the 55-foot *Ketty Lund* and the 75-foot *Sea Dog*, each morning from New Smyrna Beach to the coordinates and commenced an expanded grid search with underwater cameras and various scanning sonars. After 5 days there was still no sign of the Mariner wreckage. I quickly checked the Board of Inquiry information. This revealed something that for some reason I had never noticed before. The report by the *Gaines Mill* never said where the vessel was heading. Thus we did not know exactly off what quarter she sighted the "fireball." Despite saying the aircraft was "overhead," the *Gaines Mill* must have meant within the vicinity. Bright had agreed the wreck could be further north, and on the last day his vessels dashed to that area.

By the next Monday I was shocked when it was confirmed that nothing had been found to positively indicate the Mariner. My sitting duck in 78 feet of water had, like all other Triangle victims, vanished, even from the bottom of the sea. I was seeing disaster. My disappointment over the swamp was no longer mediated by the sure belief that the Mariner would at least be found, and that for the first time a ship or plane firmly rooted in the litany of Triangle victims would be shown to the world.

It isn't comical to see a person with their stomach kicked in, but that must be what I looked like. I decided to let the enigma of Flight 19 remain with the swamp. Is it better, after all, that it is not found? I will let the reader decide whether I have found the general last resting place of Flight 19 or not. But as for being able to put a finger on a wreck and say there it is, it will never be. Nature, the Feds, Time, everybody doesn't want the vast swamp disturbed.

All the failure and enigma aside, I did take comfort in one event coming up for which I was very proud indeed. Larry Landsman was convinced that Flight 19 needed some closure. This MS, as the reader is now aware, has revealed how much the disappearance of Flight 19 was compounded by the actions of those on the ground. Through Podesta Matoon lobbyists in

Washington DC, Larry and the SCI–FI Channel were able to lobby Congress for a Resolution honoring the men, which Florida congressman E. Clay Shaw was glad to sponsor. The Resolution was planned for late November before the documentary would air. The Gold Room at Capitol Hill was reserved for the big preview of the documentary on November 17. On that day, by invite only, members of the families were there, preparing to await the Resolution. The room was done up with a big screen at the far end and smaller ones along the side walls. The chairs were set. All waited as the Resolution was in progress. Then, suddenly, E. Clay Shaw came rushing in announcing that the Resolution passed 420 to 2! There was great applause and the premier, only 15 minutes of snippets from the actual documentary, was shown.

Strangely, however, that fickle fate that dogged Flight 19 had attached itself to me. The reasons why the men were honored was never clarified. Larry had directed an AP reporter to me so I could give the whole story of what was in this MS. I had spent 50 minutes on the phone with her. When the story broke on November 18, I had discovered the one way to screw up the AP was to tell them everything. The long article, picked up by the UP, IP, or every other P that might exist, including the British P, contained nothing of what I had said. Rather, it concluded with a quote from me, saying that I blamed the loss of the flight on the Bermuda Triangle's "electronic fog."

I about launched my breakfast. I sounded like some nut with a fringe idea. Furthermore, it was hardly clarified and sounded quite ethereal. "'It seizes an airplane,' Quasar said, 'and flies with you.'" Good gosh, it sounded quite animated and malevolent!

Some British reporters attempted to mediate the error by contacting me. Richard Luscombe, a British journalist in Miami, commissioned by the *Observer* to do a piece, saw my website article refuting the context of the AP article quote. We talked for a while on the phone and his piece, though then canceled from the *Observer*, ended up in the *South China Post*, the British newspaper for Hong Kong and English speaking South East Asia. (What, no *Siberia Times*?) However, the *London Times* also ran an article by Charles Bremner. Though he didn't contact me, he obviously surfed my site. His article was excellent, and omitted the part in the end that I said the electronic fog caused Flight 19 to vanish. He did mention it accurately; that I had said such a meteorological phenomenon does exist and has been linked with other disappearances, and that Charles Lindbergh was the first known pilot to report the phenomenon.

Not surprisingly, the press latched onto every relative they could find, but only reported those who had a caustic attitude about The Bermuda Triangle. Howell Thompson's niece, Joan Pietrucha, who must have been 1 year old at the time of the incident, was given a long piece in the *Chicago Tribune*. After my infamous misquote, the article reads: "Pietrucha will sometimes watch TV specials on Flight 19. But, she said, 'If I hear them talking about the supernatural, I just tune them out. Where's the evidence?' "— an attitude which precludes hearing the evidence, does it not?

I am digressing from the point, of course. The supernatural is not involved in Flight 19. But the format followed by many in the media does create a problem that damages research into famous mysteries. I speak of the marketing of faux extremes: the antagonist/protagonist formula of cheap fiction overlaid on fact. There is no middle ground or gray area. When it comes to popular mysteries this translates down to the conventional being the natural and anything else is the supernatural.

The Pietrucha article's headline said it all: "Marine's family scoffs at Bermuda Triangle." Even after the article quotes E. Clay Shaw saying there were "so many weird things" about the training mission and "something happened out there," Pietrucha is deferred to: "My family doesn't believe in the mystique of the Bermuda Triangle. It could have been the wind, it could have been a water spout. I just don't think it was anything creepy or weird." My comments on the electronic fog and its effects on compasses were countered in the article by Pietrucha declaring "Tons of theories, but they just don't know." Yet in the AP article she does: " 'It was pure and simple pilot error,' said Joan Pietrucha, the niece of Howell Thompson, one of the navigators on Flight 19. 'I don't believe in wacky compasses.' "

Flight 19 in general has been subjected to so much error for 60 years that the only public persona it has is the ridiculous followed by gut negative reactions. Instead of tackling some of the issues surrounding Flight 19, as I carefully documented, all news articles branched from the AP story, adding enormous amounts of error never seen before. The Pietrucha story turned her uncle, a gunner, into one of the navigators, and then allowed her to dogmatically assert "pilot error" as the cause, thus unwittingly blaming her uncle. Because of a letter from Thompson speaking about a 3 hour flight upcoming, the AP regarded that as authoritative and now Problem Nav. 1, firmly established as a 2 hour and 15 minute flight, became a 3 hour hop. Alan McElhiney, the curator of the Fort Lauderdale Naval Air Station Historical Association, was quoted as saying the crew wanted *to head East* and not West, the exact opposite of even the erroneous formula "facts" on Flight 19.

So here it came and went— the moment when I thought that some of the truth and historical fiction of Flight 19 would change places. But instead that "gorilla dust" dominated. Flight 19's families still knew nothing of the topic, though remote relatives were glad to give their views upholding the mundane. Debunking the supernatural was the common formula to follow even though no one introduced the supernatural but the debunkers, and anything but the mundane seemed to qualify for supernatural.

An attempt to look up-to-date was buried by a rehash directed by relatives, and the Resolution honoring the men was essentially negated by such statements from their remote kin insisting they made their own mistakes. A congressman and an author became fair game for the carnival of tailored quotes. Flight 19 became even more buried under the cloud of popular infotainment. "I am lost" is surely true.

More than anytime I was looking forward to putting Flight 19 behind me. It had been 13 years since I found the first tantalizing clues that they had made the Okefenokee. My research was now almost at an end. I now knew enough of the swamp to place in order the last moments of the squadron.

The final scenario is bleak. By 9 p.m. that night it was pitch black. It would be even worse over the swamp, where there is absolutely nothing to reflect light. The weather had turned stormy: dark clouds, humidity and rain. There was no moon to cast any light through broken clouds nor cause the clouds to glow overhead and illuminate the terrain. Thus, if Flight 19 was in this region, they were flying blind, remaining focused on each other only from their blinking running lights.

At this time it had been over 5 hours since they knew they were lost. They had been in flight almost 7 hours. They had no radio communication with any base, nor probably were they picking up anything over 4805 kilocycles.

As their fuel dial needle sank more toward 10 gallons, the pilots would have gotten on the intercom and prepared their crew that the time was coming close. "We're going down soon." The gunners would strap themselves into their chair in the canopy. The radioman had his fold-down chair in the tunnel.

How would this moment have affected each one? That is, of course, hard to say. One can imagine the irony going through Joe Bossi's mind. He had stayed in the Navy to fly Avengers. His words to his mother of only a couple of days ago must have haunted him. "I would have no chance to fly planes like these if I came back to Arkansas City."

Of his two young crewmen, Herman Thelander would have been the tough one. Young Burt Baluk had never had much experience like this before. His youthful face could have been quivering, his lips trembling; but he would have forced a confident smile. Thelander would have been quiet in his canopy, having his own thoughts to contend with on whether his "number was up."

For the pilots, nothing more than a droning engine would be playing on in the background; a black horizon all around; moisture from the scud clouds speeding up the windshield. There would only be vague glowing dials in each of the cockpits. The gas would be near the dreaded 10-gallon mark. Each pilot was watching that and waiting for a voice to come over the mike, saying his had hit the mark. We've all experienced watching the clock for a crucial moment to come. Finally the laborious beating of the heart stops at the staccato sound of the mike being cued. The time had come.

It's very possible that Stivers' plane would hit the mark first since it was flown once before that day and possibly not completely topped-off. That is only a guess, but his is the only TBM-1 that stands apart in that respect. Lt. James B. Smith, the officer in charge of Air Plot, said they were all topped-off at night. But he didn't have the greatest opinion of his ground crews. Let us hope they serviced each plane between day flights as well.

Stivers was forever an optimist. He had survived the worst battles of the Pacific War. His voice would be low but firm as it came over the mike to give the fateful news. The others, each in their own suspenseful eye-lock on their fuel gauges, would suddenly be jolted out of their dreadful anticipation.

Now Taylor's voice would have come over the radio. "This is it. We're not going to make it. All right, keep cool and you'll do just fine."

Did the budding priest "Yo Yo" Gruebel say a prayer for them all? Did the cocky and boyish Gallivan speak to him about faith while all this was going on? Was Gallivan mad since he was soon to get out of the Navy?

Stringy little Billy Lightfoot was all alone in the back of Gerber's Avenger, sitting in the turret looking out over nothing but ink. Thoughts no doubt skipped back and forth between his brother Eugene and his mother. Billy realized that he might go the same way. Gerber knew he was alone and no doubt called him frequently to make sure he was all right. Now he would have told him to hold on, and to get out fast and get the raft when they hit. Getting the raft out after ditching was usually a two-man job. It would be up to Billy this time. That must have been going through Billy's mind as well: slip down, unlatch, open the door, open the outer panel while the water rushes in, yank out the life raft, hope it doesn't stick, inflate it, cut the cord, wait for Gerber to walk over the wing and get in— push off!

That scenario, of course, depends on whether they knew they were over land or still thought they were over sea. If they thought they were over good old terra firma, I would think they would bail out rather than ditch. But perhaps they would deduce that much of Florida is swampy and therefore crash landing might be preferable. To confuse matters more, Taylor's voice might have come over expressing the possibility they were over Louisiana.

Thompson was an old hat, one of the Black Sheep Squadron on the old battered *Franklin*. He would remain quiet, thinking, praying, sure he'd come out like he came out of that inferno. George Paonessa would be quiet as well, thinking about his very proud and patriotic father and family. Perhaps he would be afraid of seeing himself wake up in a hospital and be crippled. Most likely he, like all the others, was rehearsing abandoning ship.

What did they think about their pilot, Captain Powers, at this moment? What indeed was he thinking? He took over and led the flight back to land being sure west was right and that west he was going. It was now hours later. Where had they gone? Had he been wrong? What indeed must have been going through his mind at this final moment of realization that they were *not* going to make it despite his certainty of direction?

The only comfort he or any of the other pilots would now have would be Charles Taylor's voice telling them how to ditch. "It's easy; I've done it three times." At this moment he would have been their teacher, coach and leader again. "Just like you're landing," he would have told them. "Just keep easing it down. Keep your neck on a pivot. Watch your altimeters and watch for the surface."

If they were over the ocean they would have descended below the cloud covering at 1,000 feet and would have seen the whitecaps. They would have a rough idea of their altitude and could ditch to a certain extent by watching the surface instead of their altimeters.

However, if over the swamp nothing would be coming into view. They would still be descending lower and lower. It would be black. Their faces would be sweating. Their eyes locked on that altimeter, darting around to see what was coming into view as they continued to descend. They might raise their goggles and quickly pinch or rub their eyes to clear them.

"A fishbowl of ink," Devlin would have thought. He was holding on, probably also thinking "snafu" and shaking his head. No more nighttime flying! He would have a clear view of the blackness outside his turret's Plexiglas. His hand would have been on the intercom ready to tell Taylor he saw the surface. Parpart would have been ensconced in the tunnel, strapped in and leaning to look out the window to see if he could see how treacherous the sea or terrain looked.

It is possible that by this time they would have had altimeter error. This event happens from changing weather and the resultant changes in pressure, causing an inaccurate reading in flight level. Depending on the amount of pressure and other factors, this could cause hundreds of feet inaccuracy in their altimeter reading. There would have been, in essence, no way for them to calculate their distance above the ground. Their altimeters may have read 300 feet when they were actually at ground zero. This error would cause a fatal miscalculation. They would be in a descending angle, believing they still had hundreds of feet to go before leveling off to belly land. Instead they would have plowed into the swamp with blinding speed and crushing force. The result would be a very temporary crater in the slushy peat. By next day it would be covered.

On the other hand, if the altimeter read opposite, that is, showed ground zero elevation when they were still 100 feet or so in the air, they would be inching their way down by Braille to an easy belly landing, that isn't near. Anticipation of striking bottom when you're sure it must be close can almost throw you into vertigo— that feeling of being on a switchback just waiting for a blinding impact. One slight mistake and they could send their planes nose first into a dive and cartwheel into pieces.

Even with a correct altimeter, at one hundred feet something unexpected would still happen in the darkness. The Okefenokee is dense with cypress and Scotch pines, which could break a plane to bits 100 feet off the ground. Some of it is so spindly, however, that a big Avenger would shear off quite a bit of the greenery and even clip a few tops off without being broken up. In utter darkness it would take a moment to realize what was happening.

"Oh, Sweet Jesus! We're over land!" one might have wailed at the last minute over his mike before throwing it aside. His plane would then rip down through the pines.

The other pilots would hesitate. What to do? There isn't

enough gas to climb altitude and bail out. Am I about to hit?

No sooner would the shrieking voice of their comrade have stopped before perhaps a momentary burst of fire shot forth from the last bit of his fuel. Perhaps another burst off the wing-tip would signal another one hit. The ground is briefly lit. Suddenly your plane shakes and rips apart, and the last you hear is your crew shouting in agony behind you.

Of the two left one may divert to a clear spot where a prairie was temporarily revealed in the dim burst of light from the last impact. The plane eases down and suddenly the ground gives way. Muck and peat splash the window. The pilot lunges forward against the force. It is forever dark now in that tannic world of the Okefenokee.

The last pilot and crew experience only a second of loneliness. The pilot is uncertain now. There is nothing to see. He is squinting at pitch-blackness that conceals a wall of death behind it.

The propeller may hit, slicing through the tops of that lousy titi. The entire plane would descend into this basket weave of brambles and be shredded, leaving behind it a scattered debris field like a retreating and defeated army.

Finally there is silence. The flight has come to an end. Finally, there is no motion. The plane settles and there is the sound of dozens of creatures slithering about. A titi landing is survivable, but the swamp would prove an unassailable enemy.

Five brave crews— fourteen brave young men— gone. No trace at sea. No trace anywhere. The heavy weather remained over the area for a few days, including heavy ground clinging fog; over more than six hundred square miles of nothing.

It is my contention this happened. This would explain what has remained unexplainable all these years. It would explain the reports of unidentified aircraft; it would explain why there was no debris at sea. The *real* fuel estimate allows it. And this is the only place over land where 5 unscheduled planes could vanish and not raise an eyebrow or leave one trace to alert anybody it happened. To hit home my point one last time, ATC from both Brunswick and Jacksonville reported 5 unidentified planes here. They were never reported advancing by other ATC controls. Thus it is around here where that flight of 5 unidentified planes ended.

Joan Powers had a disturbing nightmare at 2 a.m. December 6 about her husband and called to talk to him, discovering that he was not there. There are many who document and study "wartime psychic encounters," if you like to put it that way, noting hundreds of incidents in which family members received a strange feeling, dream or apparition at the precise moment something dreadful happened to their loved one. Is that the case here?

Curiously, the only other weird incident involved Powers' gunner's family. The brother of George Paonessa, Corporal Joseph Paonessa, received a strange telegram. "You have been misinformed about me. Am very much alive= Georgie." While one might first suspect a demented crank who liked Twainian verbiage, there were a couple of peculiarities in the telegram. One, it was sent to Joseph at his barracks at Jacksonville, Florida. George Paonessa had known that Joseph, out of all his 4 brothers who were serving in the Marine Corps, was the brother stationed closest to him in Florida. Two, it was signed "Georgie," his family nickname. His sister, Louise, sent a number of telegrams and letters to Marine Corps and Navy offices trying to determine how and why such a telegram could be sent. For her trouble she would never receive an answer. The family would reopen the case as late as 1992. Frank Paonessa even asked their local Representative to Congress, Nita Lowry, for 'Freedom of Information' on Flight 19.

Wartime psychic encounters" is an interesting concept, but it can never be proved that Captain Powers died at 2 a.m. in the swamp and the sturdy Paonessa, his radioman, continued and got out. To believe this would be to open the door on pointless conspiracy theories that require that for some unknown and bizarre reason the US Navy should wish to keep a survivor a secret. The mystery of that telegram is very real; but "Georgie" and that telegram become just another enigma in the mystery of Flight 19.

Chapter 19

"Keep a Sharp Lookout"

NOTHING ABOUT FLIGHT 19 HAS BEEN SPARED DISTORTION. Even the final official words that ended the search and thereby forever ushered it into mystery have been the subject of sensationalism and reactionary debunkery. "Sensationalists" declare that to this day the Navy officially never gave up its search for the 'Lost Squadron.' Debunkers naturally say "nonsense." Truth, as always, seems to lie in between. On December 10, 1945, Flight 19's search was halted with a communiqué containing the post scriptum: "All planes and vessels in the area keep a sharp lookout and report any pertinent information."

While the legend part of this isn't exactly true, the post scriptum is taken to heart by all those living in Florida. Any salvor, diver, beachcomber, or fisherman and boater who has seen or found old military wreckage or airplane parts would automatically think of the five Avengers and Mariner and probably actively try and verify the find. This was clearly demonstrated when five different TBMs were found only ten miles off Fort Lauderdale in 1991, or when wreckage from a military plane was found near Banana River in 1961 and again in the Keys in 1987. No matter how remote the possibility each case was checked to see if it was from the "Lost Squadron."

But finding Flight 19 does not unravel the mystery. It only locates the final resting place. The mystery exists in the beginning, not in the end. Although the beginning is shrouded by silence, it is a mute witness that can still communicate. But we have been so busy keeping a "sharp lookout" for physical wreckage that we have overlooked vital pieces of the puzzle that the beginning does indeed contain. This is something the much-maligned "legend of Flight 19" did not do. The legend of an obscured sun, of an ocean that doesn't look as it should, and wacky compasses was slowly born out of several honest attempts to try and figure out how 5 planes could vanish in an island crowded area like the Bahamas without the pilots ever seeing land.

In the mid-1970s, as bits of the legend started to crack and fall away, we learned, with no small thanks to Larry Kusche, that the Board of Investigation records still existed; and with this we learned that Flight 19 did indeed get out of the Bahamas and was last located by a general radio fix off New Smyrna Beach. In ballyhooing this discovery, debunkers of the legend overlooked the real problem the legend tried to address: the flight still got lost in an area where it was basically impossible to do so. We can dismiss Eckert's lines about the pilots not knowing 'which way is west and that the sea does not look like it should,' but we are still left with a disturbing scenario; namely, that Taylor's compasses *were* malfunctioning and 4 other pilots did not pick up on his erred course heading. There is, in fact, no logical explanation for how the flight could have gone off course for long if Taylor's compasses were the sole culprit. None of them noticed their compasses; none of them noticed the position of the sun. Why?

Something has to explain this. Eckert introduced apocryphal lines, but the true testimony of Don Poole places another question mark on this conundrum. He testified before the Board and before all and sundry the rest of his life that he told the flight to "fly west, fly toward the sun." That's not something you tell a pilot to do if the sun is not out and shining. They could see it at Fort Lauderdale; the pilots of Flight 19, too, should have been able to see it.

The words attributed to Flight 19, as William L.P. Burke proved to the Court of Corrections, are subject to interpretation. Although out of necessity he was trying to get a client off the hook, he made nonetheless a very poignant observation. "In the absence of two-way communication records, the Naval Prosecution, better known as the Board of Investigation, assembled fragmentary bits of garbled radio conversation whose authorship was only presumed, and proceeded to construct an interwoven pattern of a continuous stream of radio transmission supposedly from members of Flight 19, chiefly from FT-28. These transmission supposedly from Flight 19 could be interpreted to indicate that the pilots did not have the slightest idea of where they were, in which direction lay safety, and, in short, were dubious as to which way was up."

There are no doubts about the words of Flight 19 anymore. But what weight do they carry compared to the silence? Silence occurs more than audible dialogue. And how are we to interpret silence? The silence between Great Stirrup Cay and Taylor suddenly realizing they are lost is more significant than anything. Interpret it? We have actually ignored it.

Could Burke's summation have better applied to this period of silence more than to the period of overheard dialogue? Those silent 35 minutes from Great Stirrup Cay to "Powers, what does your compass read?" are the true mystery of Flight 19. The drama begins when Cox overhears them, but the mystery begins much earlier.

Some of the dialogue presumed to be Taylor's is arguably not his. For instance, at one point Cox asks Taylor (presumably) if he has IFF on. The response is "Negative." Yet minutes later Cox informs Fort Lauderdale that Taylor has IFF. Only one plane in Flight 19 did not have IFF. FT-36, Powers' plane, was delivered without it. Who was Cox actually talking to at one point?

Cox also insisted that he understood Taylor was calling a man named "Powers." He did not confuse Powers' name with "tower." But this seems hard to believe at first glance. If that were the case there would be no reason for Cox to want the tower informed. It wouldn't have anything to do with them. Fort Lauderdale's log actually makes more sense. It abbreviates his call to them: "It seems to be a ship or plane lost and is calling Tower x Suggest you inform Tower of it." How clearly was he truly picking up what was being initially said?

Comparing radio logs also reveals assumed authorship of dialogue. Pensacola and Houma could never pick up Powers, and this they so stated. Dialogue that exists in Port Everglades' log but not in theirs can sometimes be deduced as Powers' words. One example is when Taylor first asked "What course are we on?" "Holding course 270" is logged by Port Everglades as Taylor's dialogue, but it was not picked up by Pensacola or Houma. Taylor obviously did not ask and answer his own question. Pensacola picked up Taylor saying "Affirmative" to that, while Port Everglades did not. Therefore we can plot a quick succession of dialogue. Taylor then begins to "suggest" that they head east again.

Port Everglades' operator was clearly guessing at who was saying what. Therefore there is some reason to speculate that Powers was the one who said "I receive you very weak. We are now flying two-seven zero degrees [west]. . . We will fly two-seven-zero degrees until we hit the beach or run out of gas."

What other dialogue was never recorded into the logs? Hines says there was more. By not entering it in, have we lost not only large chunks of conversation but also rendered the tenor of what we do have obscure? How should one interpret what we do have? Robert Cox's reaction impresses upon us that he felt some urgency to help. But the transcribed dialogue doesn't indicate much panic at all. Interestingly, J.B. O'Brien, with Flight 22, recalls they were in a mild panic. Maybe it's memory contamination. Maybe it isn't. There is no doubt about it, in this Attorney Burke was quite correct: the words that we have *are* subject to interpretation.

Memories are fickle, innocently so at the best of times, and for the early drama of Flight 19 we have nothing but Cox's memory. Who said what and how much was lost? At one point in the flight, around 6 p.m., one pilot announces: "We may have to ditch any minute." But that does not come to pass. The flight is still heard an hour later. What was happening to make the pilot panic? Were his gauges erratic like the legend sometimes portrays?

However unlikely it seems, it is not impossible that some of the legend of Flight 19 is true. Inaccurately applied at the wrong stages of its drama perhaps, with some paraphrasing thrown in, but there is still mystery. Rationalizing can't really explain it all, nor can debunkers condemn our attempts to do so.

We will not know what caused Taylor's compasses to malfunction; nor what caused 4 other pilots not to notice. We won't know why the position of the sun did not alert them they were on a wrong course; nor why the ocean whitecaps did not tell them which way the wind was coming from; nor why only one pilot "thinks we should head west."

In the long run, Taylor is not to blame. He is guilty for getting confused. This confusion did not begin until they reached broken land. He was not confused when he called Powers. He merely admitted they were lost. His lost sense of where he was is hard to explain, but he was not irrational. He did in fact keep that flight together. Powers isn't to blame either. He took over and headed them west, but they seem to have ended up north and then nowhere.

Don Poole is not the villain many thoughtless people have made him out to be, including one speaker at the Fort Lauderdale Historical Association memorial service who was so mean to him Poole refused to come back. If anything, he had a better command of the actual flight schedule, route, and ETA, than any other officer that night. If his testimony can be construed to indicate he was covering up for other officers it may have been unavoidable. The Board may have overlooked it since it would be impossible to essentially tie a rope around all of them. Poole's inability to take quick remedial action is also remedied by one thing: who would have thought that anything like this could happen!?

The Navy's guilt is itself mediated to a certain extent by this solitary fact. The bizarre, convoluted disappearance of Flight 19 is beyond even the Navy's fault, although if Fort Lauderdale and ASR had acted quickly they probably would have brought them back. Cox very likely could have done so if Poole had let him go. Cox had reasons to be bitter with Poole, but accusing him of dereliction of duty was being unfair.

In her grief and anger Charles Taylor's mother created a barrage of accusations against the officers, some justifiably, others for being innocent bystanders. Mary Carroll did as well. They were incapable of accepting it was an accident, and applied intentional misconduct, which was never present on the part of Naval officers. She wrote an acid and vengeful letter to Lt. Commander Poole, reading in part: "I hope the loss of our boy Charles Carroll Taylor <u>haunts</u> you one-tenth as much as it haunts me. That you will live long enough to suffer in your family the loss of the one you cherish most! Only such a fate will even the score."

Ironically, Don Poole is the only member of the staff of Fort Lauderdale to still be alive, at the age of 97. He lives in his own home in a retirement community in Fort Meyers, Florida. There are no pictures on the wall of any of his children, although he had 3. I did not ask why or what happened. There are only B&W photos of himself and his late wife during his Navy days, pictures of himself and his squadron on the carrier during his Pacific tour of duty, and then pictures of stars he met later at his post as Navy Liaison to none other than Hollywood. Life seemed to stop for him after he retired from the Navy as a Captain so long ago in 1963. There is no picture that post dates this time. Twice the amount of time has passed in retirement as he spent in the Navy, but he still remembers Flight 19 and he still remembers the "hate" letters. He cockily dismisses them, but one still wonders if some of the curse of Mary Carroll has come true.

Bibliography

The reconstruction of the saga of Flight 19 cannot be found in any published source. It exists in a living bibliography of families and friends who remain today and recall the events of that night in 1945 and the lives of the men who vanished. Published works can bring to the reader the public persona of Flight 19, and they are listed below. Information from living witnesses and public vignettes are counterbalanced by official documentation, the most readily accessible being the hundreds of pages comprising of the official "report"— *Board of Investigation into the five missing TBM airplanes and one PBM airplane convened by Naval Air Advanced Training Command, NAS Jacksonville, Florida, 7 December 1945, and related correspondence.* Naval Historical Center. The following sources were used to compare and consult at one point in the author's research.

Berlitz, Charles F.   *The Bermuda Triangle*,   Doubleday, 1974.

Burgess, Robert F.   *Sinkings, Salvages, and Shipwrecks*, American Heritage Press, 1970.

Eckert, Allan W.   *The Mystery of the Lost Patrol*, American Legion, April 1962.

Kusche, Lawrence D.   *The Bermuda Triangle Mystery— Solved*, Harper & Row, 1975.

     "   "   *The Disappearance of Flight 19*, Harper & Row, 1980.

Gaddis, Vincent   *Invisible Horizons*, Chilton, 1965

      "   "   *The Deadly Bermuda Triangle*,
                                   Argosy Mag., February 1964.

Titler, Dale M.   *Wings of Mystery*, Dodd, Mead,
                                   1962, 1966.

Acknowledgements

The author wishes to thank the following persons, institutions and associations for their generous help, time and expertise in making this book possible. Mention of any of the names below does not imply their endorsement of any conclusions within the work. The author would also like to call to the attention of the reader the repositories where documents and photos might still remain, should the reader wish to continue in the path of Flight 19. The accident reports filed by Lt. Samuel Hines and the accident report covering the Powers/Belvito incident are in NARA II in College Park, Maryland. The records consist of roughly 1,000 boxes which are organized by plane type in Record Group 72, Entry 62 "General Correspondence, 1943-45, accident, casualties and salvage." There also might be relevant records in Entry 67A "Confidential General Correspondence, 1945." There are roughly 90 of these boxes pertaining to aircraft accidents. The elusive photo in the eastern prairie is also probably at NARA II in the Army Air Corp photos, Group number unknown.

Gary Anderson, Historian
Allan F. McElhiney, Naval Historian, President of Naval Air Station Fort Lauderdale Historical Association.
Linda Lunyou, Wayne County Historical Society.
Darla Gebhard, Brown County Historical Society.
Ina Campbell, President of Bloomfield Historical Society
Dorothy Johnson, Bloomfield Historical Society.
Patricia Herrera, Corpus Christi Historical Society.
Gloria Pritz, Historian.
Alice Seeley, genealogist.
Bob Singleton, Greater Astoria Historical Society

Jack Green, US Navy Historical Center, Washington Navy Yard.

Roxann Polzine, Nobles County Historical Society.

Miss Fern Thelander, sister of H.A. Thelander.

Heather Ferguson, Cherokee Strip Land Rush Museum.

Steve Quartz & Melanie Aitken, National Personnel Records Center, Military Personnel Records.

Jessica Louchheim, Associate Producer.

Bill "Smitty" Smith, TBM pilot, & wife, Rita Smith.

Douyou Graphics.

Barbara Gonzalez, niece of C.C. Taylor.

Carol Anne Rossi,   "   "

The History Channel.

Mary Beth Montgomery, sister-in-law of George Stivers, Jr.

Phyllis Anne Barry, his niece.

Susan Spengler, daughter of Edward Powers Jr.

Robert F. Gerber, brother of Forrest James Gerber.

John Bossi, brother of Joseph Tipton Bossi.

Jonathan Grupper, Producer.

David White, TBM pilot, NAS Fort Lauderdale instructor, search pilot.

Joe O'Brien, TBM pilot, NAS Fort Lauderdale pilot "Flight 22" Dec. 5, 1945.

Captain Don Poole, USN Retired, Flight Officer, NAS Fort Lauderdale.

Bruce Burgess, Producer.

Paonessa Family.

Charles L. Baluk, Jr.

Cecil Carter, Okefenokee swamper.

Rhonda Kimbrough, Program Heritage Director, US Forests in Florida.

Jack Mays, former Folkston Mayor, radio personality, journalist, historian, sage.

Sari Bick, Production Assistant.

Jim Carpenter, Sound Engineer.

Michael Minock, Director of Photography.

Chip Campbell, Okefenokee Swamp concession manager, boat pilot.

Verda Gerwick, Lexington Historical Society.

Bob Moore, TBM owner.

Pamela Bredenkamp, photographer.

Richard Black, genealogist.
David Glisson, pilot.
Alton Carter, Sage.
Austin Hickox, Mayor of Homeland, Georgia.
Bill Chessar, Okefenokee born and raised.
Jim Burkhart, Public Relations, Okefenokee Swamp.
Chris Trowell, historian.
Bob Bold, aviation historian.
Confederate Air Force
Weldon Richman
Mr. and Mrs. John Stefanelli.
Mike Stowe, wreck researcher
Larry Landsman, Special Projects Director, SFI-FI Channel
Jon Shriber, Senior Producer, NBC News Productions.
Lester Holt, NBC journalist.
Keith McKay, NBC producer.
Tim Beecham, NBC producer.
Patrick Huyhge, author, researcher.